THE BOOK
NOBODY READ

THE BOOK NOBODY READ

'One of the most astonishing and obsessive feats of scientific gumshoeing ever undertaken... Utterly fascinating'
Chicago Tribune

'This is an absolutely riveting book, rich in historical and scientific anecdote... combin[ing] the appeal of an authoritative history, a travelogue, and a detective story.'
Martin Rees

'Gingerich... teaches us a huge amount about early publishing, the rare book trade and book collecting, as well as the intellectual history that is his real purpose.'
Times Higher Education Supplement

'A fascinating story of a scholar as sleuth. His enthusiasm for what might be judged a rather fine point of history is infectious. His book deserves to be read not only by historians and bibliophiles, but by anyone with a taste for arcane detective adventures and a curiosity about the motivations of scholarly perseverance... As in most good adventure stories, the rewards are in the pursuit itself.'
New York Times

THE BOOK
NOBODY READ
Chasing the Revolutions
of Nicolaus Copernicus

Owen Gingerich

for Robert

Owen Gingerich

April 2016

arrow books

Art credits: Pages 3, 18, 23, 47, 50, 57, 103, 114, 120, 125, 155, 176, 184, and 250, photographs by Owen Gingerich. Page 24, photograph by Owen Gingerich; used by permission of the Royal Observatory, Edinburgh. Page 31, photograph by Owen Gingerich; used by permission of the Swedish Academy of Sciences, Stockholm. Pages 38, 43, 62, and 211, photographs by Charles Eames. Page 65, used by permission of Biblioteca Casanatense, Rome. Page 71, courtesy of the Prague University Library. Page 74, used by permission of Wroclaw University Library. Page 105, photograph by Owen Gingerich; used by permission of the University of Edinburgh Library. Page 107, photograph by Miriam Gingerich. Page 142, photograph by Owen Gingerich; courtesy of the Padua State Archives. Pages 145 and 198, used by permission of the National Central Library, Florence. Page 150, photograph by Owen Gingerich; used by permission of Cremona Biblioteca Statale. Page 165, courtesy of the Leipzig University Library. Page 188, courtesy of the Bayerische Staatsbbliothek. Page 215, National Art Library the Victoria and Albert Museum, London. Page 234, photograph by Seruei Grib. Page 241, used by permission of the Glasgow University Library. Page 252, used by permission of the Paris Observatory.

Color plate credits: 1a, photograph by Charles Eames. 1b, photograph by Charles Eames; used by permission of Uppsala University Observatory. 2, courtesy of Strasbourg Cathedral. 3, photograph by Charles Eames. 4a, photograph by Charles Eames; used by permission of the Uppsala University Library. 4b, photograph by Owen Gingerich; used by permission of the Royal Observatory, Edinburgh. 5a and 5b, photographs by Owen Gingerich. 6, used by permission of the University of Chicago Library. 7a-f, photographs by Owen Gingerich; 7a and d used by permission of the Uppsala University Library; 7b used by permission of the University of Geneva Library, 7c courtesy of the Russian National Library, Moscow; 7e used by permission of the Basel University Library; 7f courtesy of Schaffhausen Stadtbibliothek. 8a, photograph by Eric Long. 8b, photograph by Miriam Gingerich.

First published in the United States of America in 2004 by
Walker Publishing Company Inc.

Arrow Books
The Random House Group Limited
20 Vauxhall Bridge Road, London, SW1V 2SA
www.randomhouse.co.uk

Addresses for companies within
The Random House Group Limited can be found at:
www.randomhouse.co.uk/offices.htm

The Random House Group Limited Reg. No. 954009

A CIP catalogue record for this book
is available from the British Library

ISBN 9780099476443

The Random House Group Limited supports the Forest Stewardship Council (FSC ®), the leading international forest certification organisation. Our books carrying the FSC label are printed on FSC   ® certified paper. FSC is the only forest certification scheme endorsed by the leading environmental organisations, including Greenpeace. Our paper procurement policy can be found at www.randomhouse.co.uk/environment

Printed and bound by CPI Group (UK) Ltd, Croydon, CR0 4YY

Contents

SPRING 1543. Europe was in turmoil. The German princes had taken over the banner of Protestantism from an aging Martin Luther, and Europe was poised on the brink of war. Nicolaus Copernicus was lying on his deathbed when his fellow clerics at the Frauenburg cathedral brought him a long-awaited package. From the German printer Johannes Petreius in Nuremberg, hundreds of miles away, a precious sheaf of paper had finally come to this northernmost Catholic diocese in Poland, the opening pages (but the last to be printed) of the greatest scientific book of the sixteenth century. On the first sheet stood the title: *On the Revolutions of the Heavenly Spheres*. Brother Nicolaus scarcely knew what an epoch-making treasure he held.

Fast forward by four and a half centuries. The Polish astronomer's book has become a classic of the Scientific Revolution. On my desk is a cartoon strip in which a youngster reports to an impressed parent that he is studying Copernicus' *De revolutionibus* in school. The punch line is in the second panel: "Yes, we're learning how to pronounce it!"

That in itself is no mean feat, but essential to the tale that follows. "Day-revoluty-OWN-ibus" is a good approximation. The full printed title is *De revolutionibus orbium coelestium libri sex*, literally "Six Books on the Revolutions of the Heavenly Spheres," but there is evidence that Copernicus intended only the short form, *On the Revolutions*. So *De revolutionibus* is the way it is almost invariably referred to, except very occasionally when the astronomer's name itself—"a Copernicus"—becomes a synonym.

De revolutionibus was branded "the book that nobody read" by Arthur Koestler in his best-selling history of early astronomy, *The Sleepwalkers*. Koestler's highly controversial account, published in 1959, greatly stimulated my own interest in the history of science. At the time, none of us could prove or disprove his claim about Copernicus' text. Clearly, however, Koestler, a consummate novelist famous for his gripping *Darkness at Noon*, saw the world in terms of antagonists. Creating a historical vision with Kepler as hero demanded villains, and Koestler placed Copernicus and Galileo into those roles. Copernicus became his hapless victim.

My personal connection with Copernicus, though I could not have known it then, began in the wee hours of June 20, 1946, as the SS *Stephen R. Mallory* pulled away from its dock in Newport News, Virginia, on what was to prove a troubled yet memorable voyage. The *Mallory* was a reconditioned Liberty ship, outfitted with enough stalls to hold 847 horses, destined for war-torn Poland. It was a year after the conclusion of World War II, and the United Nations Relief and Rehabilitation Administration had set up a massive assistance effort. As part of the program, they sent thousands of horses to a devastated Eastern Europe. The Brethren Service Commission, an arm of one of the historic peace churches, had its own "Heifers for Relief" project. They cut a deal with UNRRA: If the United Nations would provide ships for the heifers, they would help find cowboys for the horse ships. Thus my father, a professor of history at a small Mennonite college, rounded up thirty-two potential cowboys—most of us real greenhorns—and headed for Newport News that summer. I was the second youngest cowboy aboard that ship, just past my sixteenth birthday and hence by a narrow margin qualified for a merchant marine seaman's card. The passage of time has blurred my impressions of that ocean journey, but the desolation of Poland, with its concomitant black marketeering and prostitution, burned searing images into my memory.

Two decades later I was a newly minted Ph.D. in astrophysics with a nascent attraction toward the history of astronomy. At an international astronomical conference I met Jerzy Dobrzycki, an astronomer from Poland with similar interests. I related my earlier adventure with the horses, after which he urged me to visit his country again, in particular when the historians of science were gathering for their international congress in 1965. So I did, and I soon became deeply involved with the grand worldwide preparations to celebrate the Quinquecentennial of Nicolaus Copernicus, five hundred years after his birth in 1473, which brought me frequently to Poland.

One expectation worried me. I had become a Smithsonian professor of astronomy and history of science at Harvard, and it seemed unlikely that the Quinquecentennial would pass without my being expected to lecture about Copernicus. But following centuries of Copernican scholarship, what remained to be discovered? What fresh insights could I possibly offer during the forthcoming anniversary celebrations? And what if Koestler were really right, that *De revolutionibus* was so technical and dull that nobody read it?

The epiphany dawned when I least expected it, in the Royal Observatory, Edinburgh, where I was exploring a huge safe full of rare astronomy books in November 1970. Among the rows of volumes, I found a first edition of Copernicus' book. Here, surprisingly, was a copy richly annotated from beginning to end. If the book had so few readers, I wondered, why was one of the handful of copies I had ever examined so thoroughly studied? Curiously, the parts expounding the Sun-centered, or heliocentric, cosmology were barely marked, while marginalia abounded in the latter, heavily technical passages. Who could have done this, and what might I find if I looked at some more copies of the book?

It took some sleuthing to discover that the anonymous annotator of the Edinburgh Copernicus was the leading astronomy teacher of northern Europe in the 1540s, one Erasmus Reinhold. His book became the catalyst that inaugurated my obsession to survey every surviving copy of Copernicus' book. That quest led me hundreds of thousands of miles, from Aarhus to Beijing to Coimbra to Dublin, from Melbourne to Moscow, from St. Gallen to San Diego. And even Edinburgh had only just begun to reveal its surprises.

Koestler was, I am happy to report, quite wrong in declaring that *De revolutionibus* was the book nobody read, though it took the better part of a decade to be sure and thirty years to carefully document the book's impact. Eventually I found copies owned by saints, heretics, and scalawags, by musicians, movie stars, medicine men, and bibliomaniacs. But most interesting are the exemplars once owned and annotated by astronomers, which illuminate the long process of acceptance of the Sun-centered cosmos as a physically real description of the world. Here are fascinating turf battles among astronomers as well as the struggle of the Church to come to terms with the new reality.

What follows is the story of how an intensely technical sixteenth-century treatise launched a revolution even more profound than the Reformation and how the copies have evolved into million-dollar cultural icons. But more specifically, it is a personal memoir about the making of *An Annotated Census of Copernicus' De Revolutionibus*, a four-hundred-page reference work published in February 2002. The *Census* describes individually six hundred printed copies of Copernicus' magnum opus. I dedicated those results to the members of the Petreius Society, a wholly fictitious organization that I named after the original printer of *De revolutionibus*. To qualify, a member must have seen at least a hundred sixteenth-century copies of Copernicus' book, that is, either the 1543 first edition or the second edition, a reprint of 1566. My wife, Miriam, a companion on many of my book-hunting field trips, has shared much of the adventure and surely has met the criterion for membership in the Petreius Society. Jerzy Dobrzycki, who had invited me to return to Poland, became an invaluable confederate in the enterprise; his deep knowledge of all things Copernican and his linguistic skills provided an essential resource for the project. During the years as the census progressed he became professor and eventually director of the Institute for the History of Science of the Polish Academy of Sciences. The other member of this exclusive society is Robert S. Westman, a professor of history at the University of California at San Diego, originally a rival with his own independent search for Copernicus' book, and eventually a trusted teammate.

Were they still alive, I might have considered honorary membership for two others. One was Harrison Horblit, a leading collector of rare science books, who formerly owned the most important copy of Copernicus' book in private hands. It was a presentation copy inscribed by Copernicus' only disciple to the dean at the University of Wittenberg, and a book that figures in the story that follows. Horblit was the first such collector I met, and his enthusiasm for early books lured me into this lovely vice; my own collection now includes a number of volumes that were once his. The other was Edward Rosen, a professor of history at the City College of New York, and, as my survey began, the leading authority on Copernicus and his work. I still frequently refer to his many books and translations, even though, as my memoir makes clear, I ultimately differed on several of his firmly held opinions.

For nearly half a century I have written detailed weekly letters to my parents or to our children, and copies of this extensive correspondence have proved invaluable for establishing many otherwise forgotten details and a firm chronology for the episodes recounted here. The conversations and court testimony are based on vivid memories and, I believe, accurately reflect the historical flow of events.

Besides my immense gratitude to the members of the Petreius Society, I would like to thank Dava Sobel, Kitty Ferguson, Mark Gingerich, and Dennis Danielson for their critical readings of all or parts of earlier drafts of this memoir. Special thanks go to Edward Tenner, who suggested I write this book. But most of all, I wish to register my appreciation for the enthusiastic reception of this project given by George Gibson, publisher of Walker & Company, and my admiration for his perceptive editing.

Finally, I must add a few words about the illustrations. I have taken the majority of photographs found here, and I greatly appreciate the privileges extended to me by librarians and collectors throughout the world. Several of the pictures were taken by the eminent designer and photographer Charles Eames, and I thank his estate for permission to use them. The color reproduction of Tobias Stimmer's portrait of Copernicus on the astronomical clock in the Strasbourg cathedral turned out to be a special project. The portrait is twenty feet above the floor level, and use of the picture required the permissions of the mayor of Strasbourg, the antiquities commission, the archdeacon of the cathedral, and the president of the region. I thank Georges Frick and William Shea for their effective diplomacy. And last, I owe a debt of gratitude to Teasel Muir-Harmony for her dedication in processing the images for reproduction.

Cambridge, Massachusetts,
July, 2003

THE BOOK
NOBODY READ

Chapter 1

A DAY IN COURT

"DO YOU AFFIRM to tell the truth, the whole truth, and nothing but the truth?"

I had never been at a jury trial before, much less in a witness box. My conservative religious upbringing had made me wary of such occasions, and especially the swearing of oaths. The judge, in agreement with the FBI, had not merely acquiesced to my own idiosyncrasy; on that August day in 1984 all the witnesses simply affirmed their intention to tell the truth. The defendant, once a seminary student, stood accused of carrying across a state line stolen property worth more than $5,000—specifically, a copy of Nicolaus Copernicus' *De revolutionibus*.

Stolen book cases are quite rare in court. Most are settled by plea bargaining, if not before a trial begins, then immediately after the prosecution has demonstrated its seriousness by impaneling a jury. But in this case the defendant held a security clearance for his job, and to accept a plea bargain would have immediately stripped him of that essential cachet.

I had watched the process of jury selection in the Washington, D.C., federal district court with curiosity and increasing amazement. Challenges abounded. A distinguished-looking retired black police officer was summarily dismissed despite his long service on a grand jury. A grade-school librarian was challenged. Any possible appreciators of books were eliminated. It was all too clear that the defense was hoping that a jury with the least possible education would be the most sympathetic.

What happened next threw me into near panic. The defense lawyers moved that the witnesses be sequestered. I knew the word only from the newspapers: When a jury is sequestered, the jurors are locked up in hotel rooms whenever they are out of the courtroom. I did not look forward to spending the duration of the trial incarcerated in a Washington hotel. I was relieved, but also annoyed, to discover that in this case the term merely meant that the witnesses could not hear each other testify. But in the end the strategy backfired, because we independently reinforced each other's testimonies and the jury knew we hadn't been allowed to coordinate our replies to the questions.

After the opening statements, which I was not allowed to hear but which I found out about afterward, I was brought in as the first expert witness. The government attorney, Eric Marcy, began by asking me, "Who was Copernicus, and why was he important?"

I explained that Nicolaus Copernicus was a great astronomer who is sometimes considered the father of modern science. He was born in Poland in 1473, went to school in Cracow at the same time Columbus was setting sail to the New World, but did his important astronomical work in the 1500s. His great book, *De revolutionibus*, argued against the common belief that the Earth was solidly fixed in the middle of the universe. Instead, he proposed that the Sun was immovable in the middle, and that the Earth went around it, along with the other planets. In other words, he proposed the arrangement of the solar system much as we know it today. This is why his book, first published in the year he died, 1543, is such a landmark and is eagerly sought by collectors.

I was teeming with a lot of news about Copernicus, and I could have kept the jury there the rest of the morning, but before I could go farther Marcy thrust Exhibit A—a copy of *De revolutionibus* itself—into my hands. Had I ever seen this copy before?

I informed the jury that for more than a decade I had been engaged in a study of Copernicus' book, and that I had personally examined several hundred copies, looking for notes written in the margins by early owners. I quickly pointed out that the pages were originally sold as loose sheets, and that each owner had the book bound to his own tastes. Unlike modern books, which are as similar as peas in a pod, sixteenth-century books were each individually bound and distinctive. A popular binding, espe-

NICOLAVS COPERNICVS TORNÆVS BORVSSVS MATHEMAT. NAT. ANNO 1473. OB. 1543.

Nicolaus Copernicus, from the 1654 Paris edition of his biography by Pierre Gassendi. The portrait is based on the traditional self-portrait.

cially in Italy and France, was limp vellum, similar to the "sheepskins" that go into diplomas. In Germany pigskin was often used over oak boards, generally with individual patterns rolled onto the material under considerable pressure. In England, and on the Continent as well, calf bindings over heavy cardboard were popular, usually with some pattern or another of scored rectangles. I carefully examined the book as if I hadn't seen it in years (though the FBI had refreshed my memory by showing it to me a few hours earlier), and then produced a binder of typescript sheets that I had carried to the stand with me.

"This copy is bound with a paper pattern on the sides," I pointed out, "which is very unusual. It seems to match one that is currently missing from the Franklin Institute in Philadelphia. According to my notes, it was bought from Ars Ancienne, a Swiss firm specializing in rare books, and here is a penciled note reading 'AA,' which matches that. My description also mentions that an earlier stamp has been erased from the title page, and here you can see the traces. Furthermore, it looks as if two bookplates have been removed from inside the front cover. One of them is horizontal, which is rather uncommon. I just happen to have examples of the Franklin Institute bookplates with me."

With a flourish I produced the two bookplates, one vertical and the other horizontal, and demonstrated that they fit the rectangular glue marks like keys in their locks. The book was handed around to the jury.

Before I could go farther, Marcy turned to Exhibit B, a small yellow bookseller's catalog from Old Printed Word in Washington. Had I seen this before?

"Many people know I'm searching for every possible copy of Copernicus' book, and so in fact three years ago, in the summer of 1981, a friend sent me a copy of this catalog. I saw at once that it included a *De revolutionibus*. It jumped right out at me, because, while most of the books on the list cost $50 or $100, the Copernicus is offered at $8,750."

"What did you do then?" Marcy asked.

I replied that in 1971 I had made notes on the first- and second-edition copies at the Franklin Institute in Philadelphia, but when I returned four years later to recheck several details, the second edition couldn't be found.

The description in the yellow catalog seemed to match the copy I knew was missing from Philadelphia, so I called the librarian at the Franklin Institute and suggested that he get in touch with the FBI.

The jury was spared some of the details of what had happened next. Emerson Hilker, the librarian, called the Philadelphia FBI; but upon learning that the book had disappeared more than seven years before, the bureau promptly lost interest. The statute of limitations had passed and with it the possibility of a criminal suit for the theft of the book. Hilker rang back with this bad news, wondering what he should do next. "Can you make 100 percent sure that the book is ours?" he asked me.

I told him that I could call the bookstore and ask for it to be sent on approval. A few additional details might be more definitive. So I phoned the Old Printed Word bookshop to inquire about getting the book.

"I'm sorry," the proprietor, Dean Des Roches, responded when I called him. He explained that the bookshop didn't actually own the book but simply had it on consignment, so he couldn't send it on approval.

But when I learned he actually had the book in the shop, I asked him to describe the condition of the title page more exactly. He fetched the book and told me the title page had a tiny wormhole, and then he added that it looked as if an oval library stamp had been erased.

That perfectly matched my notes, which recorded both the wormhole and the oval library stamp, so I called Hilker back, saying I was now completely convinced that the book was the Franklin Institute's missing copy.

Only later did I learn from the Washington, D.C., FBI what had transpired after the call from the Franklin Institute. Mr. Hilker had simply called the Old Printed Word bookshop, declared that the book was theirs, and asked for it to be sent back. This thoroughly alarmed Des Roches, for he was already suspicious of the consignor and had wondered how the book had been obtained in the first place. On the other hand, if it had been acquired legitimately, he would have been liable for some thousands of dollars if he merely mailed the Copernicus back to Philadelphia. So he called the local FBI, explained what had happened, and mentioned that the consignor, John Blair, apparently had lots of other items in his Maryland home.

Although the original theft was safely beyond the statute of limitations, the transport of stolen goods across state lines was a federal felony and presumably more recent. If so, the clock had been set running again.

Feeling sure a felony was afoot, the FBI agents posed as book buyers and paid a visit to Mr. Blair. They confiscated hundreds of small trade catalogs, once considered almost throwaway items, but now valuable ephemera from America's early industrialization. Many still bore the stamps of the Franklin Institute Library. Of special interest was a medical manuscript written by the famous colonial doctor from Philadelphia and a signer of the Declaration of Independence, Benjamin Rush.

Furthermore, it turned out that John Blair had been an employee of the Franklin Institute. It was well known in library circles that the institute had fallen on hard times, and it was rumored that the library was in a state of considerable confusion and with low security. The trade catalogs, for example, had been tied in bundles and crowded into the aisles of the stacks, so that users simply walked over them to get to the books. Many bundles had broken open, and hundreds of catalogs were strewn randomly in the aisles. When Blair claimed that the trade pamphlets were simply being thrown out by the institute, he had a defense that was potentially highly embarrassing to the library. Nevertheless, these titles were still listed in the library catalog, with no evidence that they had been thrown out, and surely the Benjamin Rush manuscript would never have been discarded.

In the FBI's opinion, the case was clear except for one small detail: The bureau wasn't sure precisely when the Copernicus book had crossed the state line from Maryland into the District of Columbia. But when, in the opening statement, the defense lawyers had conceded that this had happened recently, the FBI agents out in the hallway were elated. It became clear that the proposed defense would take a completely different tack.

Because the defense had listed another bookseller as a witness, the prosecution had deduced that the attorney for the defense, Andrew Graham, would try to argue that the Copernicus book was worth less than $5,000, in which case the crime would be a misdemeanor and not a felony. Had the book been a first edition, that would have been no defense, since copies were then selling for around $40,000. With a second edition, as in this case, the situation was more ambiguous. I had been alerted to the possibility that the value might become an issue, so before the trial began I had called a London dealer who had a second edition on offer just to check on the current price.

Eric Marcy ended my interrogation by asking how much Exhibit A, the Copernicus book, was worth.

Graham leapt to his feet at once. "Objection! The professor is not a specialist in book pricing!"

"Objection overruled," announced the judge, who was no doubt as curious as the jury concerning what price an old book might fetch.

To give an idea of the book's value, I cited several recent auction records. At an Amsterdam sale in 1978 a second edition had fetched $6,500, and at a Munich auction three years later a copy had brought $9,000. A few months earlier I had personally sold a copy to the library of San Diego State University for $6,800. Again Graham objected, saying that the current price was irrelevant to the price a few years back (when the supposed crime had occurred), and again he was overruled. Then I mentioned that a copy currently offered in London carried a price of $12,500. The present copy was not as good as the one in London, I admitted. After all, the defense had deliberately exhibited its wormholes to the jury. Once more Graham objected on the grounds that I was unqualified to price books, and this time the judge told him to stop interrupting me. Perhaps the asking price in the Old Printed Word catalog—that is, $8,750—was a little high, I said, but it seemed generally right.

Now it was time for a cross examination. Had I been informed, asked the defense attorney, that the book had to be worth at least $5,000 for a felony trial to continue? Yes, I knew this, I said, because I had heard the charges read at the beginning of the trial.

With the air of someone about to play his trump card, Graham then asked, "When you discussed the various auctions, you didn't mention them all, did you? For instance, what about the price of only $2,200 fetched at a Sotheby's sale three years ago? And what about their sale on April 30, 1979, where the copy went for only $3,500?" He looked very pleased, as if he had just punctured my credibility.

That's true, I said, but it takes two to tango. If at a particular sale there is only one interested buyer, he can walk away with a real bargain, below the actual market value. I mentioned that I had recently seen the first copy he had cited, which was in a private collection in Italy. It had very interesting manuscript annotations, and if it were offered by a dealer, the price would have been several times higher than the auction bid. As for the other copy, it was very tattered, with pages browned and water-stained, which greatly diminished its value.

Being a competent defense lawyer making the best of a difficult situation, Graham didn't blanch but bravely carried on. "Have you ever taken a course in book appraising?"

"No, I've never taken such a course. Nor have I ever taken a course in the history of science, even though I'm Professor of the History of Science at Harvard."

"Just answer my question!" he snapped, but it was too late. Sub voce, the judge murmered, "He's trying to."

I was about to step down when the prosecutor rose for a final query. Marcy asked, "Have you ever communicated with members of the defense? Have you ever refused to help them?"

I responded that I had indeed spoken with Mr. Graham and answered his questions. He had asked me how I was sure that the book was the one missing from the Franklin Insitute, and I explained it in detail. With that I left the stand, rather awed by Marcy's quick-witted ingenuity in framing that parting shot.

Naturally I was full of curiosity as to what would happen next, but of course I was barred from the courtroom by the sequestering motion. Only later did I fill in the details from the other witnesses and from my cousin Betsy, a Washingtonian who watched the day's proceedings with much interest. Next on the stand was "Doc" Des Roches, proprietor of Old Printed Word. He had been asked some of the same questions as I had, including the one about taking a course in book appraising, and he informed the jury that there was no such thing. His estimation of the value of the book tallied with my own, and of course he had a lot more details about how the book had been brought to him on consignment and how John Blair had also brought hundreds of nineteenth-century trade catalogs for him to sell. Blair had told him that he and his father had bought the catalogs over a long period in flea markets throughout eastern Pennsylvania, but Des Roches had inquired why so many seemed to have a uniform system of library marks on them. The next batch brought in had the marks erased, and many bore the labels of "The Library of the Saturday Afternoon Club," an ephemeral private organization that, said Blair, had been abandoned. According to the FBI, it was so ephemeral that it never existed except in Blair's imagination.

In defense, Blair claimed that he had bought the Copernicus at Renninger's flea market near Lancaster, from a dealer whose name he could no longer remember, and that the trade catalogs were simply being discarded by the Franklin Institute. His story was not very persuasive. A large box of these trade catalogs was introduced as Exhibit C, part of the group confiscated by the FBI. Emerson Hilker, the Franklin Institute's librarian, explained that the numbers on these pamphlets matched the institute's system of classification, and that in fact these titles were still in its card catalog. It seemed that John Blair had worked very methodically through the bundles of trade catalogs at the Franklin Institute, selecting only the most valuable ones.

I left Washington on Tuesday evening before the trial had concluded, and over the next few days I became increasingly curious and worried when I didn't receive any word of the outcome. Finally, on Friday around noon I called the prosecutor's office, only to discover that Eric Marcy was still in the courtroom. At about 3:30 he called me back. Late Thursday afternoon the jury had seemed hopelessly deadlocked, he reported, and he had feared that a retrial would be necessary. On Friday morning the jury had asked the judge many questions and then reheard Blair's testimony via a tape recording. Some hours later they reached a verdict: guilty.

Though the judge gave Blair a suspended sentence, the judgment was devastating. The defendant immediately lost his job, and later the FBI informed me that his wife walked out on him as well. His avaricious scheme had been thwarted because he picked the wrong book to steal.

Chapter 2

THE CHASE BEGINS

GEORG JOACHIM RHETICUS and Erasmus Reinhold are far from being household names. A few specialists will recognize Rheticus as Copernicus' only disciple and Reinhold as the author of the handy tables that embodied Copernicus' astronomy. But in the mid–sixteenth century, Reinhold was the leading mathematical astronomer in Europe and the astronomy professor at Wittenberg—the hub of the German educational system—while Rheticus was the professor of mathematics there.

Wittenberg—Hamlet's university in Shakespeare's imagination—had opened in 1502 as a sleepy provincial Saxon school that enrolled about forty students a year, but within three decades it had become the seething central pivot of the Lutheran Reformation. In 1517 Martin Luther (who had been teaching at the university since 1508) had posted his ninety-five debating theses on the castle church door in Wittenberg, setting into motion the Protestant religious upheaval. Wittenberg and its university remained Luther's headquarters as he defied both pope and Holy Roman Emperor, while its students and faculty became the reformer's ardent supporters. By 1527 the Lutheran heresy had become German orthodoxy.

Luther's chief educational lieutenant, Philipp Melanchthon, guided the university. His admirers called him *Praeceptor Germaniae*—the teacher of Germany. "I am born to war," Luther wrote, "but Master Philipp walks softly and silently, tills and plants as God has gifted him richly." Melanchthon was enthusiastic about astronomy, and when Wittenberg's longtime astronomy professor died and the mathematics professor transferred to the philosophy faculty, he arranged to appoint two young graduates to the vacant posts.

Reinhold, who got the senior appointment, had come from Saalfeld, a small town midway between Leipzig and Nuremberg. He had enrolled at Wittenberg in 1530 and received his master's degree there several years later. His obvious abilities won him a faculty position almost immediately thereafter. From all accounts he was a steady, well-beloved teacher. In those days the faculty members took their turns at being dean, and Reinhold would serve in that position in 1540 and again in 1549, and for a spell in 1549–50 he was the rector. Even when he wasn't officially dean, he was sometimes asked to write the headings in the official matriculation book because his handwriting was clear and bold.

Meanwhile, Georg Joachim Rheticus had matriculated at Wittenberg in the summer of 1532 along with 130 other young scholars. Few of his classmates made enough of a mark to be listed in Christian Jöcher's *Allgemeines Gelehrten-Lexicon*, a standard source of German Renaissance biographies. But he was bright enough to be noticed by Melanchthon, and in 1536, when he was twenty-two with a fresh master's degree, he was recruited to be the lecturer in "lower mathematics"; in the same year Reinhold became the lecturer in astronomy.

Rheticus was clearly of a different stripe from Reinhold. "Rheticus" was in fact not his original name. Some of his colleagues had taken fancy Greek translations of their names. Schwarzerdt became Melanchthon, though nearly everyone simply called him Philipp. Joachim Camerarius, the humanist leader who eventually persuaded young Rheticus to transfer to Leipzig, was born Kammermeister. "Rheticus," however, was not a translation of a vernacular family name. Young Rheticus was christened Georg Joachim Iserin, but when he was fourteen, his father, a medical doctor, was convicted of swindling and beheaded. The traumatized teenager was obliged to take another name. When he enrolled at Wittenberg, he used his mother's maiden name, de Porris, but soon he became known as Rheticus, an adopted toponym, taken from the Roman province of Rhaetia surrounding what was then and now the southwestern part of Austria.

Though he was a close colleague of Reinhold, who referred to him as "our Joachim," it is not at all clear that they were good friends. Besides being immensely interested in astrology (concerning which Reinhold was rather indifferent), Rheticus became closely affiliated with a group of raucous and iconoclastic young poets. One of them scandalized the community with a vulgar lampoon of the university leaders, including a veiled insinuation that Reinhold's wife was unfaithful. The young man was sent packing, and most of the rest of the group found Wittenberg too uncomfortable for their continued presence. Even Rheticus decided it was time to take a trip, and armed with letters of introduction from Melanchthon, in 1538 he headed south to Nuremberg.

There he met the resident scholar, Johann Schöner, who busied himself with astrology, paper instruments,* and publishing the archive of important astronomical manuscripts left over from the previous century, when Johannes Regiomontanus, the most important astronomer of the 1400s, had lived there. Presumably, Rheticus found out from Schöner† about the new cosmology under development in that "far corner of the Earth" (as Copernicus himself described the northernmost diocese in Poland where he lived and worked). No doubt Schöner told him that the Polish astronomer, who worked as canon at the cathedral in Frauenburg,‡ had some incredible notions about fixing the Sun in the center of a planetary system, and throwing the seemingly solid, immobile Earth into motion. How in the world Schöner knew about Copernicus is anyone's guess. Some have assumed that since by 1514 Copernicus had sent out at least one known copy of his preliminary prospectus for his heliocentric system, the so-called *Commentariolus*, Schöner probably saw a copy. Or Schöner could have

---

*Schöner published an equatorium, a book with movable disks that could be set to determine the positions of planets. Such disks are called volvelles, and these paper instruments were sometimes used as actual calculating devices and sometimes just for pedagogy. For example, beginning in 1538 the introductory astronomy textbook used at Wittenberg, Sacrobosco's *Sphere*, always included three or four teaching volvelles, and immediately they were copied in the editions published in Venice, Paris, and Antwerp.

†Because Rheticus later dedicated to Schöner his "first account" of the heliocentric system written while visiting the Polish astronomer, we can suppose he learned about Copernicus during his visit to Schöner in Nuremberg.

‡Now Frombork in northern Poland; I will use the Polish name when referring to modern geography.

heard the news through the astronomical grapevine that demonstrably connected sixteenth-century astronomers. Witness the case of Reiner Gemma Frisius, a Dutch doctor and mathematician. As early as 1531 Gemma found out about Copernicus from a well-born Pole, Johannes Dantiscus, who had spent some time in the Low Countries and who had then served as Gemma's patron. So somehow the news had got around.

Life's exigencies had prepared Rheticus to be a rebel, and the heliocentric cosmology, so contrary to the deeply rooted beliefs of the day, must have inflamed his imagination. Psychologically wounded by the execution of his father, he was ready to thumb his nose at a conservative society scarcely prepared to entertain such a radical cosmology. Since there were apparently no details to read in Nuremberg about these stimulating novelties, Rheticus resolved to go to the source to find out precisely what Copernicus was proposing. He may well have received encouragement from both Schöner and the Nuremberg printer Johannes Petreius, who had an interest in seeing Copernicus' work published. Thus in 1539 Rheticus set out on the long journey to the shores of the Baltic in northernmost Poland. While there he decided that making a detailed map of that region would be a good idea, so it was fortunate that the young student he had recruited to go with him, Heinrich Zell from Cologne, had some experience in cartography. By what route they traveled, and by what means, is lost in the fog of history.* Did they walk? Or go on horseback? Or on a wagon? Since Rheticus on his journey to Frauenburg took along three handsomely bound large volumes for Copernicus, and probably some of his own books as well as clothes, he presumably had some form of transport. Quite possibly he used the sixteenth-century equivalent of a rental car, buying a horse in Germany and selling it after he arrived in Poland.

Rheticus must have intended a relatively short visit to Frauenburg, little imagining that his sojourn would stretch not just for a few weeks

---

*The same is true for Copernicus himself. When he was Rheticus' age, in 1498, he journeyed from Poland to Italy for his graduate studies. We have no record of precisely when he left or arrived on his 900-mile journey to Bologna, or how he traveled.

Map of the portion of Europe relevant to Copernicus' life and times.

but for several years. Copernicus had had no formal connections with academia since graduate school in Bologna and Padua. He had neither students nor colleagues who could understand the technicalities of what he was doing. Hence the arrival at the Frauenburg cathedral of a young Wittenberg mathematician would no doubt have provided a unique and even exciting opportunity for the aging Copernicus. The Polish astronomer welcomed his visitor, eager to explain the advantages of his new cosmology.

In retrospect it seems quite remarkable that the sixty-six-year-old Catholic canon at the Frauenburg cathedral—for that was Copernicus' tenured position, which gave him time and support for his astronomical studies—could take on as a long-term guest a twenty-five-year-old teacher from the centrum of Lutheranism. But in those days, before the Catholic conservatives at the Council of Trent had finally hardened the ecclesiastical lines, the Protestant struggle was still viewed as an intra-family quarrel.

For years Copernicus had been writing a treatise on his Sun-centered cosmology, as already promised long before in his *Commentariolus*, and by then he had a thick manuscript. But he had never had a disciple, someone to whom he could introduce the intricacies of his astronomy. So the two of them began to discuss his *hypotheses*, as they were called then, really a term closer in meaning to our modern word *devices*. Copernicus must have told Rheticus about both levels of his hypotheses—the big one being his cosmological arrangement that put the Sun near the middle of a system of planets (including the Earth) wheeling around it, and the secondary, more technical, batch of hypotheses that accounted for the details of planetary motion. As the two men sat together discussing the details of the heliocentric astronomy, the young Wittenberger became increasingly convinced that the world needed to learn what Copernicus had wrought.

Rheticus must have realized that there was no publisher in Poland who could take on a work so extensive and complex. It required a printer with an international outreach to make the publication financially viable. Even Wittenberg, with its busy textbook publishers, was hardly the place for such an enterprise. Maybe this is why Rheticus brought along the three bound volumes as a gift for Copernicus. Three of the five titles included therein had been printed by Petreius.* They gave visible evidence that the Nuremberg printer could handle Copernicus' magnum opus. Whose idea was this? Maybe Johann Schöner in Nuremberg had suggested that such a display could persuade Copernicus to send his manuscript back to Germany, or it might have been Petreius himself. Schöner was well connected with the Petreius shop, dusting off old manuscripts from the Nuremberg archives or producing new works of his own and sending one to press every year or so.

But Copernicus was reluctant to release his book to a printer. Scholars have deduced that he wanted time to incorporate the trigonometric methods of Regiomontanus's *Triangles*, one of the gift books from Rheticus, into the mathematical section of his treatise. Rheticus had brought along from Schöner some observations of Mercury, which Copernicus needed to upgrade the section on Mercury. And in making changes to some of the parameters in the planetary theory, he had not had time to bring the tables into full agreement with the revised numbers.† The book was still filled with inconsistencies not as yet ironed out. And Copernicus feared it would just be an object of scorn and derision, or would simply become the book nobody read.‡

To be persuasive, Rheticus needed a further strategy. He asked, and gained permission, to publish an introduction to Copernicus' astronomy. The booklet of seventy pages was printed in the spring of 1540 in nearby Gdansk. In the *Narratio prima*, or "First Report," Rheticus did not shock his readers at the outset with the heliocentric cosmology. He obliged them

---

*The three bound volumes, shown in plate 4, contained five titles. Ptolemy's Greek *Almagest* (Basel, 1538) was bound alone. Witelo's Greek *Optika* (Petreius, 1535) was bound with Apianus's *Instrumentum primi mobilis* (Petreius, 1534). The Greek edition of Euclid's geometry (Basel, 1533) was bound with Regiomontanus's *De triangulis* (Petreius, 1533).

†Since Copernicus's original manuscript survives, which is most unusual for a book printed in the Renaissance, it is possible to see that this is the case.

‡In the preface to his book, Copernicus expressed his reluctance to publish, saying he feared he would be "hissed off the stage" and: "The scorn that I had to fear on account of the newness and absurdity of my opinion almost drove me to abandon a work already undertaken."

to work through a number of pages discussing, for example, complex details of "the motion of the Sun" before springing the big surprise: "These phenomena can be explained, as my teacher shows, by a regular motion of the spherical Earth, that is, by having the Sun occupy the center of the universe while the Earth, rather than the Sun, revolves on a great circle." Rheticus then rehearsed the reasons he had found the heliocentric arrangement compelling. He summed up his arguments by declaring, "All these phenomena appear to be linked most nobly together, as by a golden chain; and each of the planets, by its position and order and every inequality of its motion, bears witness that the Earth moves and that we who dwell upon the globe of the Earth, instead of accepting its changes of position, believe that the planets wander in all sorts of motions of their own." He even added an encomium to Prussia, perhaps yet another ploy to soften Copernicus' reluctance, or possibly an attempt to secure the patronage of Duke Albrecht of Prussia.

The *Narratio prima* was received with such interest that a reprint appeared in Basel the following year; unlike the original printing, it actually displayed Rheticus' name on the title page. Still Copernicus hesitated, and Rheticus lingered. At last, after twenty-eight months in Poland,* Rheticus was entrusted with a copy of Copernicus' manuscript destined for the Petreius press in Nuremberg, and Rheticus undertook the tedious journey home to Saxony.

Having returned to Wittenberg in 1541 with the manuscript in hand, the long overdue Joachim was appointed a full professor, a signal recognition because then there were only four professors (including Reinhold) in the arts faculty. Because the curriculum was still influenced by the medieval quadrivium†—arithmetic, geometry, astronomy, and music

---

*There is recently discovered evidence that Rheticus actually returned briefly to Wittenberg in December 1540 and delivered a short course on Sacrobosco's *Sphere*. He must have informed Reinhold, Melanchthon, Schöner, and others about Copernicus' book.

†This division of the mathematical arts goes back to Pythagoras in Greek antiquity; in the fourth century A.D. the Roman encyclopedist Martianus Capella laid the foundations for the medieval curriculum based on the seven liberal arts: the introductory *trivium* (grammar, rhetoric, and logic) and the more advanced *quadrivium*.

AD CLARISSIMVM VIRVM
D. IOANNEM SCHONE-
RVM, DE LIBRIS REVOLVTIO
nũ eruditissimi viri,& Mathema
tici excellentissimi,Reuerendi
D. Doctoris Nicolai Co-
pernici Torunnæi, Ca-
nonici Varmien-
sis,per quendam
Iuuenem,Ma-
thematicæ
studio
sum
NARRATIO
PRIMA.

Georgius Joachimus Rheticus

ALCINOVS.

Ἀεὶ δ᾽ ἴσαν ἥπιον ἦναι τῇ γνώμῃ τὸν μέλλοντα φιλοσοφεῖν

Ex dono Laurentij Viver iuir Batenbrug.

D. Luurche

The title page of Rheticus' Narratio prima (Gdansk, 1540), the first printed announcement of Copernicus' heliocentric cosmology.

theory—it made sense to have two of the professorships in mathematics. Reinhold had become professor of upper mathematics, and thus the astronomy teaching devolved on him. Rheticus became the professor of lower mathematics, that is, arithmetic, geometry, and trigonometry. He arranged for a local press to print the trigonometric part of *De revolutionibus* under the title *De lateribus et angulis triangulorum* (On the Sides and Angles of Triangles), an up-to-date mathematics text since it included what was only the second published table of sines. Its title page clearly designated Copernicus as the author, even though Rheticus himself had greatly expanded the tables.

In addition to his teaching, Rheticus became dean of the arts faculty for the winter semester of 1541–42. Apparently the appointment was more of a bureaucratic chore than an honor, and faculty members rarely served two consecutive terms. Yet even before he had left Frauenburg, he had asked Duke Albrecht in Prussia (an important patron) to petition the elector of Saxony and the University of Wittenberg for another leave of absence, this time to take Copernicus' manuscript to Petreius' shop in Nuremberg and to see it through the press.

Meanwhile, Erasmus Reinhold, the senior astronomy professor who had stayed home, had edited a new, annotated edition of a traditional advanced astronomy textbook entitled *The New Theory of the Planets*, and Melanchthon, who was the quintessential preface writer, added an erudite section quoting Xenophon, Homer, Virgil, and Plato. But the most interesting front matter came from Reinhold himself, who, in his own preface, mentioned that he knew of "a modern astronomer who is exceptionally skillful, who has raised a lively expectancy in everybody; one hopes that he will restore astronomy." In case there was any doubt, in the second edition some years later Reinhold made the allusion to Copernicus specific.

Reinhold's hint and Rheticus' *Narratio prima* of 1540 had alerted the community of astronomers and astrologers that something unusual could be expected. Thus the greatest astronomy book of the sixteenth century, indeed, one of the epoch-making science books of all time, came with at least a modicum of warning. Finally, in the spring of 1543, it was ready at Petreius' press in Nuremberg. Its title page read like an optometrist's chart:

NICOLAUS CO
PERNICUS OF TORUN
ABOUT THE REVOLUTIONS OF
the Heavenly Spheres in Six Books

The first 5 percent dealt with the new Sun-centered cosmology, so "pleasing to the mind." The other 95 percent was deadly technical. It included a handbook of plane and spherical astronomy, a lengthy star catalog only slightly updated from the one in Ptolemy's *Almagest* (the "bible" of geocentric astronomy, composed around A.D. 150), detailed instructions for going from a sparse collection of observations to the parameters of the planetary orbits, and tables for the prediction of planetary positions. As the publisher's blurb, planted squarely in the middle of the title page read, "You have in this recent work, studious reader, the motion of both the fixed stars and the planets restored from ancient as well as recent observations, and outfitted with wonderful new and admirable hypotheses. You also have most expeditious tables from which you can easily compute the positions of the planets for any time. Therefore buy, read, profit." It was surely a text to be studied, but scarcely to be read straight through.

DID ANYBODY read it? This was the question that the historian of science Jerry Ravetz and I asked ourselves one Saturday evening in October 1970. We had rendezvoused in York, the cathedral town in England's largest county; I was en route to Scotland with my family, and he had come over from nearby Leeds, where he taught at the university. Ravetz and I were both friends of Copernicus because we were friends of Poland—he perhaps because he was a socialist who had retreated to England during the McCarthy era, and I felt connected to Poland because of my formative visit with the UNRRA horses twenty-five years earlier. Jerry had spent some time in Poland, where he had produced a provocative monograph entitled *Astronomy and Cosmology in the Achievement of*

Nicolaus Copernicus. When the International Union for the History and Philosophy of Science appointed a committee to plan for the forthcoming Copernican Quinquecentennial in 1973—the five hundredth anniversary of Copernicus' birth—he was a natural choice for the committee; eventually, he became its secretary, steering the committee with consummate sensitivity around issues that still aroused passions, such as whether Copernicus was Polish or German. As a historian of astronomy, I, too, had become a member of the committee, so it was appropriate that we would get together during my family's trip north from Cambridge, where I was spending a sabbatical semester.

With the celebrations coming up in just under three years, our conversation naturally turned to Copernicus and *De revolutionibus*. It was such a formidably technical book, we reasoned, that few readers could really comprehend it much beyond the opening cosmological chapters. We remembered Arthur Koestler's claim that it was the book nobody read, and, thinking about various modern "great books" programs and its inclusion in the University of Chicago's encyclopedic *Great Books of the Western World* series, we concluded there must be far more twentieth-century readers than existed in its first decades. We even ticked off the potential sixteenth-century readers who might have made it to the end.

Rheticus and Reinhold headed our list. Andreas Osiander, the Nuremberg theologian and clergyman who finished the proofreading at the press, was necessarily a reader. Then we added Johannes Kepler, the brilliant German astronomer who in 1596 wrote the first unabashedly heliocentric treatise after *De revolutionibus* itself, and Michael Maestlin, his teacher at the University of Tübingen. Tycho Brahe, the great sixteenth-century Danish observer and builder of instruments, was another obvious choice.

We paused at the name Galileo Galilei. A physicist, he had little taste for the details of celestial mechanics; we figured he might have owned the book, but that it was unlikely he would have read it to the end. (What I was eventually to learn pretty much verified this judgment.)

Another astronomer working in Italy did make the short list, however: Christopher Clavius, the astronomer behind the Gregorian calendar reform, who had specifically mentioned Copernicus in 1581 in the third edition of his introductory astronomy textbook.*

Then we added the first Copernican in England, Thomas Digges, a man who once lived in the same block as William Shakespeare and whose library might have been of use to the playwright when he was researching background material for *Hamlet*. Since Digges had translated part of Copernicus' cosmology into English, he would surely have been a reader. John Dee, an eccentric Elizabethan wizard who owned the largest private library in England, must surely have at least owned the book, even if he didn't read it all.

But with those nine likely readers we bogged down, and our conversation drifted off to other topics such as the glories of the York cathedral. Then we bid each other farewell, as early the next day my family was headed toward Edinburgh.

In Scotland serendipity took its course. The Royal Observatory in Edinburgh holds a fabulous collection of rare astronomy books formed by the Earl of Crawford late in the nineteenth century. For years these precious books had been mixed on the open shelves with the ordinary astronomy treatises, but at some point before I arrived they had been collected for safekeeping in a couple of enormous steel cupboards. Among these treasures I stumbled upon a first edition of Copernicus' *De revolutionibus* that was thoroughly annotated from beginning to end.

Such a discovery would probably have meant little to me except for that conversation just two nights earlier about how few readers we thought the book had had in its first years. If it was read so rarely, why was the very next copy I chanced upon so full of evidence of a most perceptive reader, who had marked innumerable errors and who had worked his way through to the very end, even past the obscure material on planetary latitudes that brought up the rear of the four-hundred-page volume?

---

*Although entitled *Commentary on the Sphere of Sacrobosco*, Clavius' book went so far beyond the small traditional text written in Paris by Sacrobosco around 1215 that it is considered an independent work. Until Clavius' huge expansion, in one edition or another Sacrobosco's *Sphere* had been the standard introductory astronomy text for more than three centuries.

The Royal Observatory, Edinburgh, the magnificent edifice resulting from the gift to Scotland in 1888 of Lord Crawford's astronomical instruments and library.

Furthermore, there was a fascinating motto penned across the title page (in Latin): "The axiom of astronomy: Celestial motions are circular and uniform or composed of circular and uniform parts" (plate 4b). I would have expected something like, "This crazy book fixes the Sun and throws the Earth into dizzying motion." But no such thing. Here was a reader who ignored the Big Hypothesis, but who was enthusiastic about the secondary ones. The rich annotations verified that interest—hardly anything in the cosmological chapters, but a dense thicket of marginal comments whenever Copernicus grappled with his little epicyclets that allowed him to eliminate what he believed to be one of Ptolemy's most obnoxious devices.*

---

*Ptolemy's astronomical handbook, or *Almagest*, is the classical formulation of geocentric astronomy. It is an epoch-making volume because it showed for the first time that the complex appearances of planetary motion could be accounted for by a group of relatively simple mathematical devices. But one of them, the so-called equant, was heavily criticized in the Middle Ages because it appeared to violate the celestial principle of uniform circular motion. More details, including an example of an epicyclet, are found in appendix 1 of this volume.

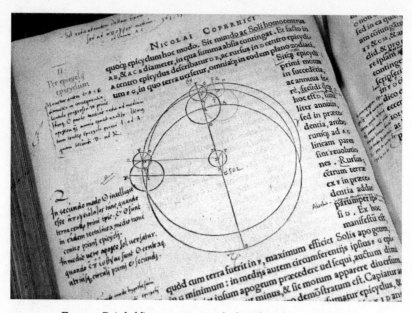

Erasmus Reinhold's annotations in the heavily technical section of
De revolutionibus, *folio 91 verso.*

Suddenly, inspiration struck. Perhaps this was a copy from one of
those nine readers! But no, that would be an unbelievable coincidence,
I told myself. There had to be at least a hundred extant copies (as I
naively thought then, badly underestimating the number), and that
would make the chance of finding a copy from one of those nine readers
about one out of ten, or less. That I had just turned up one of those
copies seemed a hypothesis too improbable to entertain for long. But
whose copy was it?

I searched in vain for an owner's name. The manuscript inscriptions
at the beginning and end provided nary a clue. Then I looked more
closely at the heavy pigskin binding. Later I would learn that it was a typical example of a blind-stamped binding—"blind" because the impressed
designs had no color or gilding. Around the edges were long patterned

strips with biblical figures. Below an empty central panel was the date 1543, and above the panel I noticed the initials *ER*.

I reacted with a shock. Could the initials stand for Erasmus Reinhold, the leading mathematical astronomer in the generation after Copernicus, and one of the astronomers on our short list of annotators? I seized a pencil and paper to make a rubbing of the dim impression and, to my dismay, found not two, but three, initials: *ERS*. It seemed my hypothesis had just evaporated.

Back in Cambridge it took a few days to sort out the significance of what I had found. I soon discovered that those three initials, *ERS*, were exactly what was required for Erasmus Reinholdus Salveldiensis, for in the sixteenth century a man's birthplace—in this case Saalfeld—was a part of his formal designation. Furthermore, the blind-stamps on the binding had been recorded and could be attributed to a Saxon binder, quite possibly working in Wittenberg. I started making inquiries to obtain samples of Reinhold's handwriting, specimens that eventually confirmed my original deduction. And I pondered the implications of the finding. If one book displayed such insights as to how a major professor of astronomy passed over the heliocentric cosmology but accepted the epicyclets, what would more copies reveal?

AND SO MY great Copernicus chase began. At first, I poked around Cambridge, Oxford, and London, all rich repositories of Copernican first editions. Oxford and Cambridge provided a special challenge because there were altogether sixty colleges, each with its own library. Cambridge was easy to search because there existed the published "Adams catalog" of sixteenth-century books in the Cambridge libraries; it took several months, however, to discover that it wasn't complete. Trinity College, Isaac Newton's alma mater and the wealthiest college in Cambridge, actually owned three copies of the first edition, whereas the Adams catalog listed only two. It seemed reasonable to the librarian to declare the third one redundant, so that copy hadn't been listed. Eventually, the fellows at Trinity realized that they faced a dilemma. Clearly they didn't need a third copy of the rare

Copernicus first edition, but it was so valuable that if it were auctioned at Sotheby's, there would be a furor in the press to the effect that Trinity College was selling off its patrimony. So the fellows quietly gave up any intention of auctioning it.*

The two first editions Trinity always intended to keep turned out to be the most interesting of the seven copies I found in Cambridge. One, originally owned by a Basel bookseller who had collaborated with the printer of the second edition of *De revolutionibus*, had already found its way to England by 1570. It was heavily and studiously annotated, probably by the Oxford scholar Edward Hindmarsh, who eventually willed his books to Trinity College, Oxford. How the book got from Oxford to Cambridge is a ruefully poignant story. It was bought in a bookshop in 1794 by one Stephen Street, rector of Trayford in Sussex, who penned the following notice inside the front cover of the book: "This may very probably be a copy of the first edition, if it be, it is worth many Guineas. . . . I hope I shall not be taken for a thief, as I bought this volume of Messrs White's House and have pasted their bill of Parcells into it." However, Street's aging father probably thought he was returning his son's copy to its rightful owner when he gave it to Trinity College, Cambridge, instead of Trinity College, Oxford.

Trinity's second copy of the first edition had been acquired in 1843, on the 300th anniversary of the book's publication, as a gift from Richard Sheepshanks, an English astronomer and fellow of Trinity. The volume was clearly censored, something that puzzled me at the time, and it had some early and seemingly minor notes concentrated in

---

*Perhaps that was just as well, because there are always unexpected dangers in deaccessioning. An extra copy of Descartes' *Geometry*, which Trinity had set aside for sale, turned out to have Isaac Newton's critical remarks scattered throughout the margins. The most spectacular example involved not Cambridge, however, but Göttingen University in Germany. The university library had once owned two copies of Newton's *Principia*, so they decided to sell the dirty one, all marked up by some previous owner. Only after the duplicate was released did someone discover that the critic who messed up the pages was none other than Newton's rival, Gottfried Wilhelm Leibniz. That copy of the *Principia* is now one of the great treasures of the Bodmeriana Collection in the outskirts of Geneva.

the final part of the book, something that perhaps should have puzzled me at the time but did not.

Finding the books in Oxford was a little harder than in Cambridge because there was no published guide. Very quickly, however, I discovered a private index in the Bodleian Library, which led me to the four first editions scattered throughout the college libraries plus one in the Bodleian itself.

Besides Oxford and Cambridge, the opening rounds of my survey included London, where there is an amazing number of libraries. Who would have guessed that "Dr. Williams's Library," rich in theological volumes, would boast a first-edition Copernicus? How I found out about that one I can no longer remember, nor do I recall who told me about the Polish Institute Library, with its partially annotated but almost hopelessly dilapidated copy. The most memorable experience, however, was not examining the first edition in University College, London, but the path to that library.

In my college philosophy class I had heard about the early-nineteenth-century Utilitarian philosopher Jeremy Bentham, but I had never expected to meet him. After he died, in 1832, in accordance with his will, his body was dissected in the presence of his friends, and his skeleton was then decked out in his clothes and seated upright in a glass-enclosed case that could be wheeled into the council room so that he could continue to participate in college affairs. The mummified head was replaced with a wax effigy. And there he sat in the front hallway of University College, his glassy eyes staring straight ahead, a half-amused expression on his waxen face. It was a distinctly memorable encounter, perhaps to make up for the library's lightly browned, eminently forgettable, unannotated first-edition *De revolutionibus*.

In those early weeks of searching for first-edition copies of Copernicus' book, I could scarcely have imagined the ultimate scope of the search. Then I was merely checking the copies to see how many contained evidence of serious readership, aiming to make a fairly simple list of locations together with brief information concerning the extent of annotations, if any. Of the dozen and a half copies I managed to locate in Cambridge, Oxford, and London, only one was thoroughly annotated, and two others had some marginalia of note. It was already apparent that Reinhold's copy in Edinburgh had been an incredibly lucky starting point, and it seemed that the yield of richly marked up copies would not be high. Perhaps it was true that the book had not many, indeed, hardly any, readers.

Chapter 3

IN THE STEPS OF COPERNICUS

WHAT SORT OF person was Copernicus? Did he like puns? Did he ever play jokes on his classmates or his fellow canons? Did he enjoy music? He probably never tasted a potato, or chocolate, or drank a cup of coffee, food that had scarcely whetted European palates in his time, but was he keen on beer, or had he developed a taste for wine in his Italian sojourn during his graduate school days?

Was he tall, dark, and handsome? Did he ever have a girlfriend? Did he like children? Alas, these are unanswerable questions. No personal memoirs exist. Of his seventeen surviving letters, fifteen deal essentially with cathedral matters, the sixteenth concerns currency reform, and the seventeenth is a long technical astronomical account, but unfortunately not dealing with his cosmology. Half a century after Copernicus' death, a professor at Cracow began assembling materials for a biography, but that story was never written and the data have been lost.

As I began my quest for copies of his book, Copernicus himself was a shadowy personality for me. His piercing eyes look out from the portrait preserved in the town hall in Toruń, his birthplace, his pupils reflecting the Gothic windows of his homeland, and his red jerkin more engaging than the drab habit of a friar. Yet he was hardly more than a cardboard figure propped up in a shop window. Among many things, during the course of my investigations these initial impressions gradually transformed into an understanding of the man and his impact.

A few weeks after my fateful trip to Edinburgh in 1970, my travels took me to Uppsala, seat of the most distinguished university in Sweden, and there the transformation began, when I first laid eyes on Copernicus' working library, the actual volumes that he had used and annotated.

The Uppsala astronomers had invited me to give several technical lectures on my own astrophysical researches. From a previous visit I knew the observatory had a magnificent library, so I made sure that my schedule included plenty of time to work my way through the shelves, book by book, handling not only great landmark volumes, such as first editions of Isaac Newton's *Principia* and John Napier's *Logarithmia*, but the many minor and often rarer works that make up the fabric of normal science. The observatory library had an outstanding showing of Kepler's titles, and, remarkably, there was even a marvelous volume that contained notes in Copernicus' own hand. On the illustrated pages of eclipse predictions in Johann Stoeffler's *Calendarium Romanum magnum*, Copernicus had recorded his observations made in the 1530s and early 1540s (plate 1b).

That volume in the observatory library was just the tip of the iceberg as far as the Uppsala University Copernicana were concerned. A century after Copernicus' death, the library of the cathedral where Copernicus had written the principal parts of his *De revolutionibus* had been captured by the Swedes during the Thirty Years' War and shipped off to Scandinavia. The great majority of these books had ended up in the University Library in Uppsala, where they had been systematically ferreted out by several generations of visiting scholars from Poland. At that moment the librarian from Copernicus' hometown, Toruń, was working in the main library, and on his desk I caught my first glimpse of the Frauenburg hoard. From him I obtained a list of books that Copernicus used, which I added to my own voluminous notes on the observatory's collection.* But I did not then have the time to inspect the Copernican books.

I would twice return to Sweden in my search for Copernicus, but then I needed to head south to Copernicus' Polish homeland, to join the international committee as it finalized the plans for the forthcoming Quinquecentennial in 1973. As secretary, Jerry Ravetz guided the proceedings,

---

*The Frauenburg cathedral library must have owned a copy of Copernicus' *De revolutionibus*, but no such copy has been located.

One of three sixteenth-century copies of Copernicus' Commentariolus, *preserved in a printed copy of* De revolutionibus *at the Swedish Academy of Sciences in Stockholm.*

reviewing the progress on a volume of studies on the reception of Copernican astronomy, and the plans for an excursion called "In the Footsteps of Copernicus."

"Some of the most important things at conferences happen on the excursion buses," my friend and fellow astronomer Jerzy Dobrzycki remarked during a break in our sessions. "If it hadn't been for a conversation in a bus at the international congress here in 1965, we might never have discovered the third manuscript of Copernicus' *Commentariolus*."

The *Commentariolus*, or "Little Commentary," documented an early stage of Copernicus' work. Never printed during his lifetime, it was apparently distributed by manuscript to a few of his confidants. For a long time this document remained out of sight to Copernican scholars, and then around 1880 a Swedish scholar discovered a manuscript copy at the

Academy of Sciences in Stockholm. A few years later a second manuscript turned up in the National Library in Vienna. At first it was dated to the 1530s, the decade before the publication of *De revolutionibus* in 1543. Then, however, researchers found an inventory of the library of a sixteenth-century Cracow professor, one Matthew of Miechow, with an entry, "A manuscript of six leaves in which the author asserts that the Earth moves while the Sun stands still." When scholars noticed that it described the document found in Stockholm and Vienna, they realized that the *Commentariolus* itself had to antedate May 1514, the date of Matthew's inventory. In other words, the "Little Commentary" showed a preliminary form of Copernicus' approach to the heliocentric arrangement, dating at least thirty years before *De revolutionibus* was published, and it offered a different arrangement of the small secondary circles than he finally adopted in his magnum opus.

Neither the Stockholm nor the Vienna manuscript was in Copernicus' hand. They were secondary copies, and they had certain discordances, not to mention that the one in Vienna lacked several leaves. Thus, finding another early copy was a discovery of some importance, and a bus ride had provided the catalyst. During a conference excursion from Warsaw to Cracow in 1965, Dobrzycki had had a chance conversation with the Scottish scholar W. P. D. Wightman. Wightman had described some curious annotations on interleaved pages of a copy of *De revolutionibus* found at his home university in Aberdeen. The annotations had been made by Duncan Liddel, a sixteenth-century Scot who had taught at Rostock in northern Europe before returning to Aberdeen with a rich collection of continental books. From Wightman's partial description of the notes, Dobrzycki guessed that they might comprise another copy of the *Commentariolus*, a hunch that turned out to be correct. It stands as the single most significant piece of Copernican research in the 1960s.

The opportunity to discuss with Dobrzycki my own recent findings in Scotland made attendance at that 1970 committee meeting particularly memorable. He was by then rapidly becoming a key authority on Copernicus, and he listened attentively to the news about Reinhold's well-annotated copy.

"If you're going to make a survey of copies of Copernicus' book, you should consider including the second edition as well as the first," he suggested. "Duncan Liddel copied the *Commentariolus* into a second edition. And Tycho Brahe annotated a second edition, an important copy that's now in Prague. Since the second edition was published in 1566, long before scholars generally accepted the Copernican system, you might double your chances for finding interesting annotations by including it."

Dobrzycki didn't have to work very hard to convince me of the merits of his suggestion, because in the Uppsala observatory library I had just seen a second edition containing an early manuscript copy of Copernicus' "Letter against Werner." The letter, written by Copernicus in 1524, is tantamount to a review of a book published in Nuremberg two years earlier and written by the mathematician Johann Werner. It is a long and quite technical essay, dealing with what was then called the motion of the eighth sphere and is today called precession of the equinoxes, and it is Copernicus' only known letter that deals with astronomy—although not at all with his heliocentric cosmology. Only four other sixteenth-century copies of the letter are known. Clearly, if material of this importance was to be found in copies of the second edition of *De revolutionibus*, they should be examined as well.

I would have immediately taken up Dobrzycki's idea to include second editions in my survey except that time was too short in Warsaw to follow up on it. I did look at the only first edition there, at the University Library. It seemed rather anomalous that the Polish capital had only a single example of the 1543 edition, but the National Library had lost its copy of this national treasure in the wanton destruction of its collection in the closing days of World War II. In any event, the University Library copy turned out to be heavily annotated, including citations to Johannes Kepler and to the French astronomer Pierre Gassendi, so clearly the notes came from the seventeenth century. Only much later did I appreciate that major annotations seldom came singly, and eventually I found that a copy of the second edition in Toronto had marginalia closely related to those in the Warsaw copy.

I DIDN'T GET BACK to Poland until the summer of 1972. By then the quinquecentennial preparations were in their final stages, and we went through a dress rehearsal of the excursion to the Copernican locales: Lidzbark, Olsztyn, Frombork, and Toruń.

Copernicus' uncle, Lucas Watzenrode, had in 1489 become bishop of Varmia, the northernmost Catholic diocese in Poland, and he provided for his nephews (Nicholas and his older brother, Andrew)* by having them appointed canons at the cathedral in Frauenburg, the religious capital of the diocese. The younger Copernicus had spent the years immediately after the appointment as a graduate student in Italy—in fact, he had actually formalized the appointment while he was in Bologna. There he had studied both civil and church law, suitable topics for a future cathedral administrator, but he consolidated his interest in the stars by rooming at the home of the professor of astronomy. During a second sojourn in Italy, right after the first, he studied medicine in Padua at the venerable university renowned for its medical instruction, but ultimately he took his doctorate in canon (or church) law at the nearby university in Ferrara.†

Following his return from Italy, from 1503 until 1510, he lived primarily with his uncle at the Bishop's Palace in Lidzbark. The tall, severely Gothic brick palace still impresses visitors. Resplendent in the sunlight, it proved to be a photogenic monument, though without the Copernican touches that we had found earlier at Olsztyn. Still, this may well have been the setting where Copernicus made the decision to turn his back on ecclesiastical politics, to forgo the obvious opportunity to become his uncle's successor, and to turn increasingly serious attention to astronomy. In 1511 Copernicus established his primary residence in Frauenburg, but our tour stopped first in Olsztyn, twenty-five miles south of Lidzbark and fifty miles southeast of Frombork.

Copernicus lived at the Olsztyn castle for about three years, between 1516 and 1519, and was there off and on after he began a second term in 1520 as administrator of the chapter's lands in this southern part of the

---

*The Latin spellings of their names are Nicolaus and Andreas. I generally refer to the astronomer as Nicolaus except when discussing his adolescent and student days.

†Scholars speculate that frugal Copernicus took his degree at Ferrara, where he had not actually studied, because he knew no one there and hence was spared the expenses of a lavish graduation party.

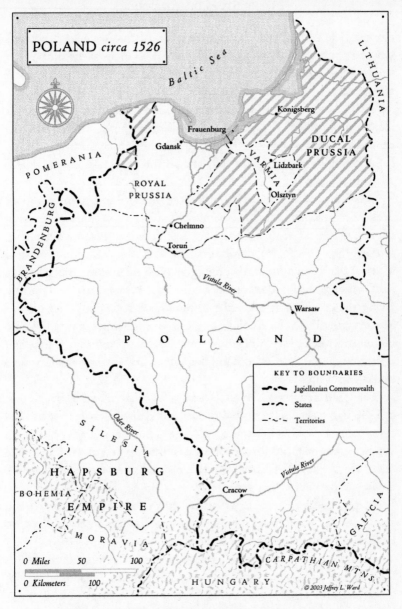

POLAND *circa 1526*

Baltic Sea

LITHUANIA

Konigsberg

Frauenburg

DUCAL
PRUSSIA

Gdansk

POMERANIA

VARMIA

Lidzbark

ROYAL
PRUSSIA

Olsztyn

BRANDENBURG

Chelmno

Torun

Vistula River

Warsaw

P O L A N D

KEY TO BOUNDARIES

— ·· — ·· Jagiellonian Commonwealth

— · — · States

—   ·   — Territories

Oder River

SILESIA

Vistula River

HAPSBURG

BOHEMIA

Cracow

EMPIRE

GALICIA

MORAVIA

0 Miles   50   100

CARPATHIAN MTNS.

0 Kilometers   100

HUNGARY

© 2003 Jeffrey L. Ward

*Varmia, the northernmost diocese in Poland, surrounded by the
lands of the Teutonic knights.*

diocese. He had already been a member of the Varmian Cathedral Chapter for nearly twenty years when they appointed him administrator; it was one of the most important posts he held as a cathedral canon. He traveled through the chapter's estates and more than a hundred villages, collecting revenues and administering justice. But from our touristic viewpoint, his most significant accomplishment here was the construction of a reflective sundial, arranged so that the light of the Sun reflected from a small mirror traced its daily and annual paths high along the wall of the castle porch.

The climax of our inspection tour came when we arrived in Frombork itself, the cathedral complex where Copernicus spent most of his adult life and where he wrote *De revolutionibus*. A monument in Polish brick Gothic, the cathedral stands within an enclosure on a bluff overlooking a bay of the Vistula River. Beginning in 1514, soon after he had written his *Commentariolus*, the astronomer rented quarters in a tower of the wall facing the cathedral facade. His tower stands about fifty feet high— kitchen and dining space below, a bedroom and living room in the middle, and a well-lighted workroom on the top story. From this height overlooking the cathedral and its compound, Copernicus pored over his books and manuscripts.

There were copies of *De revolutionibus* in Olsztyn and in Frombork, but the tight rehearsal schedule of the tour didn't allow me time to see them then. We had to hasten on, reaching Copernicus' Toruń birthplace late in the evening, and we had a brief opportunity to wander through the charming old city of Toruń after dusk had fallen. In the fading light we could see the impact the anniversary was already having on the town. Fresh paint and new construction mitigated the standard iron-curtain shabbiness. Zdenek Horský, a colleague from Prague, observed, "It's too bad every Polish town can't have such an occasion!"

As soon as the Copernican committee finished its tour, I headed to Cracow, the historic old university town in southern Poland, where ten historians from eight countries had been invited to organize an international multivolume *General History of Astronomy*. The instigator of the project, Eugeniez Rybka, was a professor of astronomy there, but I had taken the lead in actually assembling an organizing committee and bringing it to Cracow.

The visit gave us an opportunity to see the old Collegium Maius where Copernicus had been an undergraduate in the 1490s. Its splendid medieval-looking classroom, with its wooden benches and frescoed geometric diagrams, conveyed the spirit of the place where Copernicus might have studied the quadrivium. The frescoes came from a later period; but despite that, I was moved by the lingering ethos of that space. And upstairs the Collegium Maius boasts a splendid collection of early brass instruments from Copernicus' lifetime, though they arrived there a few years after Copernicus had left for his graduate studies in Italy. The jewel of the collection is a brass terrestrial globe, the earliest to show America—an elegant reminder that Columbus and Vespucci were contemporaries of the young Copernicus.

Of all the Copernicana preserved in Cracow, the most precious and most significant is the actual manuscript of *De revolutionibus*. When Rheticus finally convinced his teacher that the manuscript should be published, a copy was made for the printer while Copernicus himself retained the original working document, the one in which he had frequently sliced out obsolete pages and replaced them with new ones. Copernicus must have continued working with the manuscript, because an errata sheet was issued with the printed book, and the same errors are marked in his hand in the manuscript. A reasonable explanation for this would be that Copernicus read the proof sheets of the book, and when he caught errors there he went back to his manuscript and corrected them on the master copy.

After Copernicus' death, the original manuscript was conveyed to Rheticus, and after he died in 1574, it was inherited by his student Valentin Otto. A century later the manuscript came into the hands of the famous Danzig observer Johannes Hevelius, but after that it dropped out of sight until 1840, when Copernican scholars became aware of its existence in a private library in Prague. After World War II, this treasure was lent by Czechoslovakia to Poland, at which time the Poles simply appropriated it for themselves and deposited it in the Jagiellonian Library of Copernicus' alma mater. It would have been unseemly for one commu-

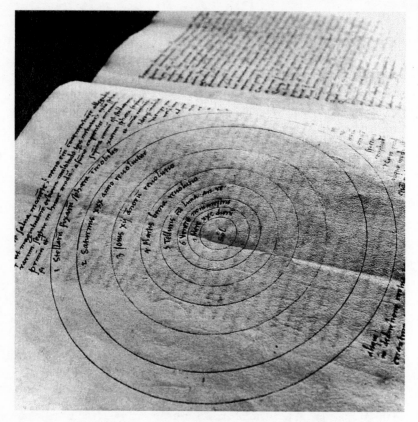

Copernicus' holograph manuscript for De revolutionibus, *photographed by Charles Eames in the Jagiellonian Library in Cracow.*

nist society to object too strenuously to a bordering brother society, so the precious document has remained in Poland.

As part of the cultural enrichment for the *General History* committee, we all paid a visit to the manuscript department of the Jagiellonian Library, where we carefully compared the manuscript with the facsimile that had just been issued by the Copernicus Commission of the Polish Academy of Sciences. The facsimile came in a normal library form, and also in a deluxe edition with a replica of the actual binding and ragged edges to the pages, each hand-cut to match the original manuscript sheets. This latter form of the facsimile was so authentic looking that on a later occasion several of my colleagues were momentarily fooled when they saw it exhibited in a museum case.*

Joining our group of historians, who had come to look at the original manuscript of *De revolutionibus*, was the designer Charles Eames, camera in hand. I had first met Eames at a lunch in the Harvard Faculty Club. At the time I knew only that he was the designer of the famous Eames chair with its tall molded plywood, black leather cushions, and accompanying five-footed ottoman footrest. During the lunch I quickly discovered that he had designed the multiscreen show for the IBM pavilion at the 1964 New York World's Fair. Shortly thereafter I began consulting with Eames for an IBM exhibition on the history of computers, an assignment that frequently took me to his design shop in Venice, California. A year or two after the computer history wall was completed for the IBM headquarters in New York City, I suggested to him that he should design something for the forthcoming Copernican anniversary. Eames swiftly agreed and persuaded his IBM patrons to put on a Copernican exhibition in their Madison Avenue display room, which is what brought him to Cracow.

After the *General History* organizing meeting was over, I joined Eames for a photographic excursion through that ancient university town. His perceptive eye caught the textures of wood and stone around the old market square and St. Mary's Church with its majestic carved altarpiece, the masterpiece of the sixteenth-century Schwabian woodcarver Veit Stoss (or Wit Stwosz as the Poles regard him)—a treasure that had just been installed when Copernicus arrived in Cracow as an undergraduate. By and by, Eames remarked that he would like to return to the library to photograph some details of Copernicus' manuscript. I

---

*Some years after the Copernican Quinquecentennial I got an earnest phone call from a Chicago book dealer asking for an evaluation of the deluxe facsimile, which was by then out of print and worth some hundreds of dollars. It didn't take long for me to guess what had happened. A well-meaning but relatively unsophisticated Polish-American citizen of Chicago was presenting a facsimile to the Adler Planetarium, and he wanted to take a tax deduction. The only problem was that he thought he was giving the planetarium Copernicus' original manuscript.

was not at all sure that we could get permission, especially because it was technically vacation time at the library, but I was willing to try.

"How many pictures does he want to take?" a long-faced manuscript librarian asked me. Knowing Charles' professional approach to photography and that he was apt to shoot three or four rolls if given a chance, I was cagey. "About ten openings," I replied with a certain amount of apprehension. That would give him an unlimited number of exposures of a selected number of pages.

With permission granted, I selected seven openings to be shot, starting with the famous heliocentric diagram, and Charles came in and set up his master-and-slave flash unit, which had no connecting wires. The librarians were so intrigued by this technological magic that they hardly noticed when he took nearly two hundred frames of the seven openings, and they even suggested another page, where Copernicus had left an inky thumbprint in the margin.

The next morning we started out at the crack of dawn, cameras in hand, for a few final hours of filming in Cracow. Then, after the morning shoot, we packed ourselves up for a trip to see Copernicus' library in Uppsala, which I had already fleetingly glimpsed on my trip to Sweden a year and a half earlier. I'll never forget Eames's astonishment when we arrived at the Arlanda airport north of Stockholm. Looking around at the waiting room, he said, with both surprise and obvious pleasure, "These are all our chairs!"

I had come to Sweden armed with the results of the Polish researchers, namely, a list of the shelf marks identifying the Frauenburg cathedral's volumes. At the University Library it took me the better part of a morning just to fill out all the call slips so that the books Copernicus actually owned or used could be fetched from their scattered locations.

It's hard to recreate the awe and tingle of excitement I felt when at last I had the assembled array in hand. Here was Copernicus' copy of Ptolemy's *Almagest*, and there were the tables he had had bound in Cracow, with a scattering of his working notes, faint hints about his creative process. And here was his Greek dictionary, with his name in Greek penned on the flyleaf. Copernicus had bought the volume while a graduate student in Italy around 1500, and he used it to translate the "rustic, moral and amatory" letters of an obscure Byzantine epistolographer, one Theophylactus Simocatta. This printed book, now extremely rare, was published in Cracow in 1509. Modern scholars have criticized Copernicus for a fairly pedestrian approach strongly limited by the inadequacy of his dictionary; his effort scarcely stands comparison with the eminent Italian translators. But from another viewpoint, it was one of the first such attempts at translation in the transalpine world, where the humanistic Renaissance had come more slowly.

Most thrilling of all were the three folios, uniformly bound in white pigskin, brought and signed over to "my teacher, Copernicus" by the young itinerant astronomer Georg Joachim Rheticus (plate 4a). Here was the Greek edition of Ptolemy's *Almagest* that had been recently published in Basel, but perhaps even more cunningly, there were three books published by Johannes Petreius in Nuremberg. A vision of Rheticus as publisher's agent sprang into mind. Often publishers' representatives had called at my office, bringing samples of their wares in the hopes that I might sign up as a textbook author. It took no great leap of imagination to see Rheticus handing over the handsome volumes of Regiomontanus' *De triangulis*, Apianus' *Instrumentum primi mobilis*, and Witelo's *Optika*, all beautifully crafted in Nuremberg, with the not-so-subtle implication that Petreius' press would be exactly the right place for Copernicus' own work to be printed.

Eames and I took the books into an open reading room and lined them up on a long table. Occasional patches of softly filtered sunlight highlighted the row of volumes. We worked intently with our Nikons, capturing the textures and patterns of those fifteenth- and sixteenth-century treasures. They are probably the best photographs ever made for recording the spirit of that evocative collection (plates 5a and 5b).

Working with these books was for me the climax of a ten-day metamorphosis that transformed my mental picture of Copernicus from a vague, dark, medieval figure to a three-dimensional human being. Was it seeing the room and parapets at the Frombork cathedral where the astronomer had lived and studied that made Copernicus come alive? Or the classroom in Cracow's Collegium Maius? Was it the vision of the young Rheticus with his gift of books gently cajoling the aging and reluctant old master into publishing his work? Or was it the manuscript notes tucked here and there in his books of astronomical tables, precious hints concerning the almost lost trail to heliocentrism?

In any event, reassembling his library in Uppsala and placing his books in the reading room's flickering sunlight at last convinced me, at a visceral level, that Copernicus really had existed, flesh and blood. I no longer had any trouble envisioning him as a real personality who had lived, dreamed, and even burned the midnight oil as he sweated over his geometry and his calculations. These very tomes sat on his desk as he struggled, pen in hand, to penetrate their ancient wisdom—knowledge that was to become the foundation for his reformation of astronomy.

IN THE MONTHS that followed I made repeated trips to the Eames office in Venice, California, to plan for the quinquecentennial exhibition that Charles would set up in the street-floor lobby of the IBM headquarters in New York City. Even pedestrians on Madison Avenue could at least partially view the labyrinth of Copernican panels that Charles and his assistants envisioned.

"It would be nice to include some original artifacts," he remarked on one of my visits, and by very early the next morning, before the staff had arrived (for I was still on eastern time), I had arranged to borrow a large astrolabe, a sixteenth-century brass sundial bowl, a first-edition *De revolutionibus*, and even one of the five copies of the 1540 *Narratio prima* in the United States.* I agreed to lend a few of my own books as well, recognizing that they projected a kind of verisimilitude and connectivity that bridged the centuries. "Using Xerox copies is like kissing your wife through a pane of glass" is the way one of my colleagues expressed it. There is a palpable linkage to the Renaissance itself when handling these antique books, and especially when a scholar from ages past has added his own impressions in the margins.

---

*The University of Louisville generously lent its *Narratio prima*. The other four American copies of this edition were at Harvard, Yale, the Burndy Library (now the Smithsonian's Dibner Library), and in the private collection of Robert Honeyman in California. Since that time the Honeyman copy has been auctioned and has become the only located copy in Italy. Meanwhile two previously unrecorded copies have come to America, one to a private collection and one to the Linda Hall Library in Kansas City.

The nave of the Frombork (Frauenburg) cathedral, photographed by Charles Eames for the IBM Copernican quinquecentennial exhibition.

Besides the artifacts, Eames was keen to convey the Copernican environment through richly textured photographs, and he realized that Frombork was missing from his repertoire. So he was determined to return to Poland, to photograph the northern countryside in the ambered light of autumn. "If you're going to the cathedral," the staff and I implored him, "you've got to get a good tall picture of its nave." So Charles flew off to Warsaw, hired a car and driver, and went photographing in northern Poland. He visited Toruń and took a beautiful photograph of the Copernican portrait that hangs in the Town Hall (plate 3). And I especially admired his image of the Frombork cathedral nave—he took along a special architect's camera just so that he could get an undistorted view of the high Gothic arches.

But I admired even more what Eames achieved on the flight back home. He began thinking very hard about how the Copernican Sun-centered model for planetary motions differed from the traditional Ptolemaic Earth-centered scheme. Each arrangement required two circles to explain the motion of Mars as seen from the Earth. In the Copernican model the Earth's orbit and Mars's orbit each circled the Sun. But in the geocentric setup the Earth was at rest in the center, with a large deferent circle (i.e., the carrying circle) going around it. The second circle (called the epicycle) contained the planet Mars and rode on the deferent. Charles knew that each arrangement had to give the same results because each system had to represent the same observations, and he wanted to make a dynamic model to show it. By the time he flew into Los Angeles, he had the device all sketched out, complete with the behind-the-scenes linkages (accomplished with bicycle chains). The front face of his model looked like this, with the straight rods representing the observational line of sight from the Earth to Mars in each arrangement:

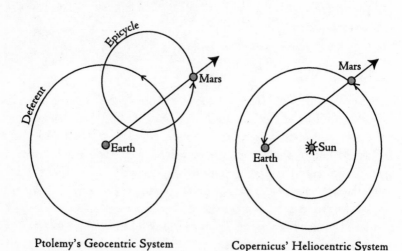

Ptolemy's Geocentric System   Copernicus' Heliocentric System

When he got to Venice, he turned the plans over to his shop techni-
cian, and a few days later he had a working model. In the exhibition,
which opened in December of 1972, the Eames machine ran continu-
ously without default for something like six months (plate 1a). As the
circles turned, the rods, representing the observed line of sight to Mars,
always remained parallel. Each time Mars came on the inner side of the
epicycle, the combined counterclockwise motions of the deferent and
epicycle caused the geocentric rod to briefly swing clockwise, the so-
called retrograde motion. Whenever that happened, in the heliocentric
model the faster-moving Earth was always nearest Mars and bypassing
it, so the heliocentric rod remained in perfect tandem with the geocentric rod. It was a brilliant demonstration of the equivalence of the two
systems, and what worked for Mars would work for each of the other
planets. Over the years several of my students managed to simulate
Eames' model on their computer screens, but no one was ever able to
figure out how to do it with bicycle chains.

Chapter 4

THE LENTEN PRETZEL
AND THE EPICYCLES MYTH

QUIETLY SLEEPING on the title page of *De revolutionibus* is a Greek epigram reading *Ageometretos medeis eisito* (Let no one untutored in geometry enter here), by legend the motto on the gate of Plato's academy in ancient Athens. Any potential buyer who could decipher the Greek would most likely have been able to handle the geometry. Nevertheless, I discovered a fair number of copies with simple annotations in the opening cosmological chapters, only to have them peter out as the going became more mathematical. In the early chapters Copernicus gave his strongest arguments for a Sun-centered blueprint for the planetary system—arguments based on simplicity, harmony, and aesthetics, for in those days before the invention of the telescope it was impossible to find satisfying observational proofs for the mobility of the Earth. Like Ptolemy's system, Copernicus' heliocentric model gave tolerably good predictions, and many astronomers were fascinated just to see another way to get more or less the same predictions. As the Jesuit astronomer Christopher Clavius at the Vatican's Collegio Romano remarked, Copernicus simply showed that Ptolemy's arrangement was not the only way to do it. But any astronomer who wanted to be sure how it worked in detail had to be well tutored in geometry.

In the decade before I became thoroughly enmeshed with Copernicus, I had already entered the magic circle of Johannes Kepler, the German astronomer who was born almost exactly a century after his Polish forerunner. Kepler's Quadricentennial came in 1971, just two years before the Copernican Quinquecentennial. I sometimes say, with considerable truth, that my career in astrophysics was derailed by anniversaries.

Johannes Kepler burning the midnight oil as he contemplated the dome required to finish the Temple of Urania, from the frontispiece of his Rudolphine Tables *(Ulm, 1627).*

Kepler, the man who really forged the Copernican system as we know it, had been the hero of Arthur Koestler's *The Sleepwalkers*. Some critics said his book should have been entitled *The Sleepwalker*, because he had only one good example of a scientist groping in the dark, Kepler himself. A few even more perceptive critics said he had zero examples, an opinion borne out today by recent scholarship. When Kepler wrote in his *Astronomia nova* about finding the elliptical shape of Mars's orbit, saying, "it was as if I had awakened from sleep," readers had little reason to suspect that the account was anything other than straightforward autobiography. They had no way of knowing that it was actually a highly structured rewriting of his personal history of discovery, designed to persuade his readers to abandon the perfect orbital circles, an idea almost as radical as heliocentrism had been a few generations earlier. But Kepler was not sleepwalking at all; the manuscripts show that he knew what he was doing, and even seemingly blind alleys brought grist to his mill. Kepler knew that he had access, through his mentor, Tycho Brahe, to observations far more accurate than any available to Ptolemy or Copernicus, and these showed that the earlier computational methods were by no means adequate, particularly for the planet Mars. New wine required new bottles.

For all its faults, Koestler's book was, for me, a fascinating read, and highly stimulating. When I read it in 1959, I was about to become heavily involved in astrophysical computing, exploiting the newly available power of the electronic computer revolution. Several years later, in looking around for a historical computing problem as sort of a busman's holiday, hopefully with a Keplerian connection, I encountered an intriguing statement in the early part of the *Astronomia nova*, where Kepler wrote, "Dear reader, if you are tired by this tedious procedure, take pity on me, for I carried it out at least 70 times." Kepler was trying to find an eccentric circular orbit for Mars based on four observations, and since he could not find a direct answer, he sought the answer through a series of iterations. Computers are particularly good at repetitive problems, and this seemed like a perfect kind of demonstration. So I programmed Kepler's geometry for the Smithsonian Observatory's IBM 7094, fed in his observational data, and the computer solved the problem in eight seconds with nine tries, the minimum required. Today's computers would solve it in a split second, but in 1964 eight seconds seemed like lightning, and the computer magazines just loved this result.

But I was left with a disturbing puzzle. If the IBM 7094 could solve the problem with the minimum of nine tries, why did it take Kepler at least seventy attempts? Did he make so many numerical errors that the iterations simply failed to converge? One way to find out, I knew, would be to examine the manuscript record. Unlike the situation with respect to Copernicus, where, apart from the manuscript of *De revolutionibus* itself, very few manuscript research materials survive, for Kepler there is a huge and only partially mined archive. For the most part, Kepler's manuscript legacy is today found in St. Petersburg, Russia. A modest amount of library research disclosed that the relevant pages might be found in volume 14 of the Kepler papers in the Academy of Sciences Archives in what was then still Leningrad. Thus, in 1965 I requested a microfilm of the manuscripts in that volume. Once every six months for five years I repeated my request to the Soviet authorities and to the Russian astronomers I had met at international astronomy meetings. Finally, in 1970, I actually got the microfilm, and a superb film it was. The archivists had disbound the volume so that it was possible to photograph the complete pages, with nothing lost in the central gutter of the binding.

Armed with the microfilm, I soon learned a lesson that I've been obliged to relearn several times. One can, in the absence of evidence, reconstruct rationally with clear hindsight how a discovery might have taken place. But Kepler was treading where no investigator had gone before, toiling with the murky, ambiguous realities of cutting-edge science, and what really happened was quite distinct from a tidy rational reconstruction. It is true that Kepler was prone to make numerical errors, but this was not his basic problem. He had, at that point in his researches, gained access to the groundbreaking observations of the great and noble Danish observer Tycho Brahe. Kepler could see that neither Ptolemy nor Copernicus was able to predict positions to a high degree of accuracy. Compared to anything available to his predecessors, Brahe's legacy was overwhelming. Nevertheless, Kepler could rarely find precisely the right observations he needed, so he was obliged to interpolate from other observations made around the same time. This process itself led to errors, and Kepler had to carry out multiple iterations just to find out where the discrepancies arose. Examining this process, and looking more broadly at the manuscript material, enabled me to warn my colleagues, during the quadricentennial proceedings in 1971, that, contrary to the received opinion, Kepler's *Astronomia nova* was far from being a simple, linear, autobiographical account of how he had arrived at his conclusions about the planet Mars.

In those years leading up to the time when I received the Leningrad microfilm, I became increasing intrigued by the technical contents of Kepler's *Astronomia nova*, which ranks alongside Copernicus' *De revolutionibus* and Newton's *Principia* in the trilogy of foundation works for the astronomical revolution of the sixteenth and seventeenth centuries. Remarkably enough, unlike the *Revolutions* and the *Mathematical Principles of Natural Philosophy,* Kepler's *New Astronomy* had never been translated into English. Aided by two well-trained classics students, I resolved to remedy this hiatus. And that is how I encountered Kepler's Lenten pretzel.

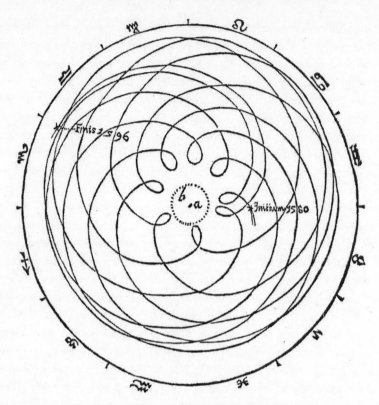

*Kepler's Lenten pretzel for the geocentric path of Mars, from chapter 1
of his* Astronomia nova *(Prague, 1609).*

Kepler's *Astronomia nova* was subtitled *Commentary on the Planet Mars*;
he opened his commentary with the observational problem he would attempt to solve, namely, to account for the appearances of Mars as seen from
the Earth, and this he illustrated with a carefully made diagram showing
how Mars tracked with respect to the Earth between the years 1580 and
1596. His diagram, shown above, was an astronaut's-eye view as seen from
far above the Earth and above the plane of Mars's orbit. Mars repeatedly
approaches the Earth (shown fixed in the center at point *a*), makes a back-

ward loop, only to recede and repeat the process roughly two years later at a spot in the zodiac about fifty degrees to the east of the previous loop. The loops themselves trace around the entire sky in approximately seventeen years, during which time Mars itself circumnavigates the sky eight times.

In his Latin text Kepler said that he was first inclined to liken his diagram to a ball of yarn, but he thought the better of that and preferred instead to call it a *panis quadragesimalis*. I recognized *panis* as meaning "bread," but what to make of *quadragesimalis*? A fortieth? Forty times? The word isn't in any Latin dictionary, but I remembered some advice given to me by a Harvard mentor, Professor I. B. Cohen, who suggested that when you are stuck on a technical Renaissance Latin word, try the second edition of *Webster's* unabridged dictionary (which, unlike the third edition, still includes many obsolete words) or the *Oxford English Dictionary*. And there it was: "belonging to the period of Lent; Lenten." This, in turn, led to an investigation into the history of pretzels.

At that time my wife, Miriam, and I had acquired an *Encyclopaedia Britannica*, and part of the salesman's pitch was that if the set did not include an answer to any reasonable question, their research team would investigate. I felt an obligation to send them questions with some regularity, as I assumed this would give employment to impoverished University of Chicago graduate students. The only time I felt fully satisfied by an answer was when I inquired about the history of pretzels. The *Britannica*'s investigator reported that pretzels had their origin in southern Germany— Kepler territory—as Lenten favors for children.

Kepler used his Lenten pretzel diagram as the starting point to show how various cosmological models accounted for this convoluted geocentric pattern. It was Claudius Ptolemy, working in Hellenistic Alexandria around the year A.D. 150, who first showed that a relatively simple geometric model could account for the seemingly complex movements of Mars and of the other planets. As the Eames machine showed, he accomplished this with two circles, a smaller planet-bearing circle that rode upon a larger deferent circle.

A careful inspection of Kepler's complex Lenten pretzel reveals that the loops differ from one another not only in how close they come to the

Earth but in their width and in their spacing. Ptolemy added two more features to his model to take these aspects into account. First, he moved the center of the deferent circle away from the Earth, to the position marked *b* in Kepler's pretzel—that could account for the fact that on one side the loops don't come as close as on the other side. This off-center position of the deferent circle gave it an alternative name: an eccentric circle. Ptolemy could not see the loops from above, however, so he had to deduce that this was happening just from the projection of these effects onto the sky.

Second, he had to figure out a way to make the epicycle move around the eccentric (deferent) circle more slowly on the side where the loops didn't come as close to the Earth, and here he invented a very ingenious device called the equant. The equant point *E* is shown in the diagram below. The angular motion is uniform about that point, so that the epicycle moves from *A* to *B* in the same time that it takes to go from *C* to *D*, because the angles at *E* are identical. Of course the epicycle had to travel faster in the segment *CD* than in *AB* because the length of the arc is greater.

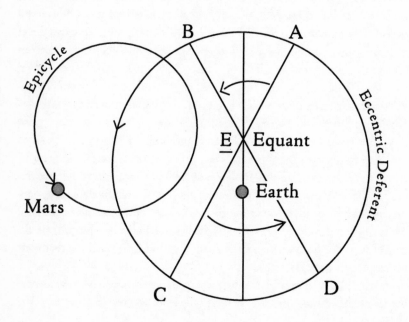

The equant got Ptolemy into a lot of trouble as far as many of his successors were concerned. It wasn't that his model didn't predict the angular positions satisfactorily. Rather, the equant forced the epicycle to move nonuniformly around the deferent circle, and that was somehow seen as a deviation from the pure principle of uniform circular motion. Ptolemy himself was apologetic about it, but he used it because it generated the motion that was observed in the heavens. Altogether his system was admirably simple considering the apparent complexity and variety of the retrograde loops.

THE MOTTO THAT Erasmus Reinhold wrote on the title page of his *De revolutionibus*, "Celestial motion is circular and uniform, or composed of circular and uniform parts," was clearly a potshot at Ptolemy and an accolade for Copernicus. Today we admire Copernicus for having the audacity to introduce the heliocentric cosmology into Western culture, essentially triggering the Scientific Revolution. The Copernican cosmology did not just provide the modern blueprint for the solar system. It was a compelling unification of the disparate elements of the heavenly spheres. The greatest of scientists have been unifiers, men who found connections that had never before been perceived. Isaac Newton destroyed the dichotomy between celestial and terrestrial motions, forging a common set of laws that applied to the Earth and sky alike. James Clerk Maxwell connected electricity and magnetism, and showed that light was electromagnetic radiation. Charles Darwin envisioned how all living organisms were related through common descent. Albert Einstein tore asunder the separation between matter and energy, linking them through his famous $E = mc^2$ equation.

Copernicus, too, was nothing if not a unifier. In the Ptolemaic astronomy each planet was a separate entity. True, they could be stacked one after another, producing a system of sorts, but their motions were each independent. The result, Copernicus wrote, was like a monster composed of spare parts, a head from here, the feet from there, the arms from yet another creature. Each planet in Ptolemy's system had a main circle and a subsidiary circle, the so-called epicycle. Mars with its epicycle was a prototype for each of the other planets, but because the frequency and size of the retrograde was different for each planet, an epicycle with an individual size and period was required for each planet. Copernicus discovered that he could eliminate one circle from each set by combining them all into a unified system. Just as the Eames machine demonstrated that a heliocentric orbit for the Earth and for Mars could give the same results as a geocentric Martian deferent and epicycle, the same could be done for each planet. If all the models could be scaled so that the Earth's orbit was always the same size, then they could all be stacked together with a single Earth-orbit, thereby reducing the total number of circles. And when Copernicus did this, something almost magical happened. Mercury, the swiftest planet, circled closer to the Sun than any other planet. Lethargic Saturn, then the most distant planet yet identified, circled farthest from the Sun, and the other planets fell into place in between, arranged in distance by their periods of revolution.

"In no other way do we find a wonderful commensurability and a sure harmonious connection between the size of the orbit and the planet's period," Copernicus declared in the most soaring cosmological passage in his entire book. What Copernicus had achieved was a linked system in which all the distances were locked into place relative to a common measure, the Earth-Sun distance, which provided the yardstick for the entire system.*

But Reinhold and his many followers admired Copernicus for a quite different aesthetic idea, the elimination of the equant. Copernicus devoted the great majority of *De revolutionibus* to showing how this could be done. While he had eliminated all of Ptolemy's major epicycles, merging them all into the Earth's orbit, he then introduced a series of little

---

*Surprisingly, Copernicus' planetary system was more compact than the carefully nested pieces of the Ptolemaic arrangement. Yet his cosmos was vastly larger than Ptolemy's because he was obliged to place the stars themselves far enough away so that the motion of the Earth around the Sun would not show any obvious changes in the positions of the stars. "So vast, without any question, is the divine handwork of the Almighty Creator," Copernicus concluded. It was a giant step for humankind, in the right direction, but ultimately that sentence made the Catholic censors very nervous. Probably, they had no objections to the vastness, but they just weren't convinced that Copernicus could know how God did it. Catholics were ordered to delete that sentence from their copies.

epicyclets to replace the equant, one per planet.* Because this made the motion uniform in each Copernican circle, the anti-equant aesthetic was satisfied. My Copernican census eventually helped to establish that the majority of sixteenth-century astronomers thought eliminating the equant was Copernicus' big achievement, because it satisfied the ancient aesthetic principle that eternal celestial motions should be uniform and circular or compounded of uniform and circular parts.

JOHANN STOEFFLER'S *Ephemeridum opus* of 1532 was one of the first rare books I had bought for my own library. I found it on the shelf of Blackwell's rare book department in Oxford. In the days when a typical scholarly book cost maybe $10, it was quite a plunge to spend $170 for a book of numbers, but it was exciting to have such an old volume for my very own. Before that I had used a similar volume in Harvard's Houghton Library to help date a poem by the satirical Tudor poet John Skelton, so I had an idea about what I was getting.† Stoeffler's ephemerides were filled with daily positions of the Sun and planets for the years 1532 through 1551.

I was particularly curious about the basis for Stoeffler's numbers because of a popular legend found in many secondary sources. According to that story, a principal reason why Copernicus had sought to create a heliocentric cosmology was that Ptolemy's hoary system had become so encumbered with jury-rigged embroideries that it was at the point of

---

*For the mathematically curious, details of Copernicus' procedure are given in appendix 1. In the *Commentariolus* Copernicus used a double epicyclet for Venus, Mars, Jupiter, and Saturn, but in *De revolutionibus* only a single epicyclet with an eccentric orbit. The Earth and Mercury had somewhat more complicated configurations—it remained for Kepler to construct a truly unified Copernican system.

†Skelton's "Garland of Laurel" included the lines

    Arectyng my syght to the Zodyake, . . .

    When Mars retrogradant reversyd his bak . . .

    And whan Lucina plenarly did shyne,

    Scorpione ascending degrees twyse nyne.

Using the Regiomontanus *Ephemerides* for 1495, I could see that on 8 May 1495, the Moon was full in the eighteenth degree of Scorpio and in conjunction with the retrograding Mars.

collapse. Astronomers throughout the ages had added one epicycle to another as they attempted to keep up with the observed deficiencies of the system.

Probably the legend had begun soon after the modern recovery around 1880 of Copernicus' *Commentariolus*, or "Little Commentary." After describing the complexities of planetary motion, Copernicus closed this account with an exclamation: "Behold! Only 34 circles are required to explain the entire structure of the universe and the dance of the planets!" Superficially, the passage looks as if Copernicus were crowing about the great simplification his system afforded. If Copernicus could handle this in only 34 circles, Ptolemy (or at least his medieval successors) must have required many more.

There is a wonderful, very old, but no doubt apocryphal, story that Alfonso the Great, looking over the shoulders of his astronomers who were compiling the *Alfonsine Tables*, remarked that if he had been around at Creation, he could have given the Good Lord some hints. The obvious interpretation was that King Alfonso's astronomers, in order to take care of the observed discrepancies between the Ptolemaic predictions and where the planets actually were, had been obliged to add more circles, small epicycles on epicycles. It's rather reminiscent of the lines paraphrasing Jonathan Swift:

> Great fleas have little fleas
> upon their backs to bite 'em
> And little fleas have lesser fleas
> and so ad infinitum.

The legend reached its apotheosis when the 1969 *Encyclopaedia Britannica* announced that, by the time of King Alfonso, *each planet* required 40 to 60 epicycles! The article concluded, "After surviving more than a millennium, the Ptolemaic system failed; its geometrical clockwork had become unbelievably cumbersome and without satisfactory improvements in its effectiveness." When I challenged them, the *Britannica* editors replied lamely that the author of the article was no longer living, and they hadn't

Johann Stoeffler from his Ephemeridum opus *(Tübingen, 1531). At least a century ago this image became confused as a portrait of Copernicus.*

the faintest idea if or where any evidence for the epicycles on epicycles could be found.

In those early years of the space age, the Smithsonian Observatory's computer spent most of its time tracking satellites. In its spare time, it calculated the flow of photons through the outer layers of stars—that was my specialty—and in my own spare time I had the machine calculate medieval planetary tables. I recomputed the *Alfonsine Tables* and discovered to my surprise that they were pure Ptolemaic, totally lacking any embroideries at all. Then, using the hundreds of cards the keypunchers had produced for me, I generated a section of the Stoeffler ephemerides. Again a surprise! My pure and simple *Alfonsine Tables* calculations closely matched the positions that the Tübingen astronomer had published in his book. His were the best ephemerides of the day, and they showed absolutely no evidence of epicycles on epicycles. A much-repeated and well-entrenched myth had just bit the dust, or so I thought.

During 1973, the great quinquecentennial year, I had occasion to mention my conclusions at a symposium in Copernicus' birthplace, Toruń. In the audience was Edward Rosen, the world's foremost Copernican scholar. A professor at City College in New York, he had built his career of fastidious scholarship on finely honed translations and the ability to dig out relevant details from incredibly obscure sources. As part of his doctoral dissertation, Rosen had translated Copernicus' *Commentariolus*, and he was particularly fond of the line about the entire ballet of the planets being accomplished in just 34 circles; he firmly believed that part of Copernicus' achievement was to simplify an overwrought system. "How can you be so sure there weren't epicycles on epicycles?" he demanded to know. Certainly, I hadn't inspected *all* the possible medieval manuscripts!

I'm not sure I ever convinced him about epicycles on epicycles, but today I understand the problem a lot better. The entire calculational procedure for the *Alfonsine Tables* depends on a clever approximation invented by Ptolemy to handle a single epicycle on an eccentric circle. Frankly, there was no mathematician in the Middle Ages ingenious enough to have devised a similarly economical computational scheme for multiple epicycles. It's not even necessary to inspect all the medieval astronomical manuscripts to be sure.

Edward Rosen had simply misread Copernicus' intentions in writing that the entire ballet of the planets was accomplished in only 34 circles. Copernicus must have realized that with his small epicyclets he actually had more circles than the Ptolemaic computational scheme used in the *Alfonsine Tables* or for the Stoeffler ephemerides. His exuberant conclusion to the *Commentariolus* surely registered his delight that, although the celestial appearances seem very complicated, a great many phenomena can be modeled with only 34 circles. There was clearly no comparison intended with his predecessors.

THE EPICYCLES-ON-EPICYCLES legend fails in yet another way. There are virtually no records of systematic observations to find possible discrepancies between where the tables predicted the planets to be and where they really were. Yet there is one minor but extraordinarily significant exception that I discovered when Charles Eames and I were photographing the Copernican books in Uppsala.

Bound in the back of his printed copy of the *Alfonsine Tables* are sixteen extra leaves on which Copernicus added carefully written tables and miscellaneous notes. Below the record of two observations made in Bologna in 1500, there is, in another ink, a cryptic undated remark in abbreviated Latin: "Mars surpasses the numbers by more than two degrees. Saturn is surpassed by the numbers by one and a half degrees" (plate 7a)."

In 1504, soon after Copernicus had returned to Poland from Italy with his newly acquired doctorate, the two slowest-moving naked-eye planets put on a splendid show as faster-moving Jupiter bypassed the slower-moving Saturn. This great conjunction, which takes place only every twenty years, provided a sensitive test of the tables, since it did not require any elaborate instruments to determine on which night the planets actually passed each other. And this time Mars joined the ballet. Between October 1503 and March 1504, swifter Mars passed Jupiter and then Saturn, and then, going into retrograde motion, went back past Saturn and Jupiter, and finally in direct motion bypassed Jupiter and Saturn yet again. It was a great celestial display, and surely Copernicus would not have missed it.

With my computer programs I could calculate where the *Alfonsine Tables* placed these planets not just during this period but for several decades of the sixteenth century, and I could compare these calculations with modern ones showing where the planets really were. To my amazement, the calculations gave a unique error signature for February and March in 1504. During that interval the predictions for Jupiter were excellent, but Saturn lagged behind the tables by a degree and a half while Mars went ahead of the predictions by nearly two degrees. Only during this interval did the errors match Copernicus' notes, so the evidence is firm that he had observed the cosmic dance at this time and was fully aware of the discrepancies in the tables. But what is most astonishing is that Copernicus never mentioned his observation, and his own tables made no improvement in tracking these conjunctions.

It is pretty clear that neither Copernicus nor his predecessors were interested in adding extra circles just to make the predictions work a little better. Nevertheless, the legend of epicycles on epicycles has become so pervasive that barely a year passes without some author in the *Physical Review* or the *Astronomical Journal* remarking, apologetically, "Maybe my theory has too many epicycles." Clearly, I haven't stamped out the myth.

Chapter 5

"EMBELLISHED BY
A DISTINGUISHED MAN"

THE COUNTDOWN for the Copernican Quinquecentennial built toward a climax in December of 1972. Charles Eames wanted to open his exhibition in the IBM building on Madison Avenue in time for the crowd of Christmas shoppers, so I flew to New York on several occasions to help with the labyrinthine assembly of the panels and artifacts. Edward Rosen came up from City College to lend his critical eye to proofreading the captions. He corrected a few minor errors but overall was quite pleased with the comprehensive presentation.

As the installation reached its final stage, the staff assigned me the task of keeping Charles occupied so that he would no longer make schedule-breaking improvements in the show. To create a dramatic Christmas showpiece on the street side of the exhibition, he had enlarged to mural proportions a page from Julius Schiller's idiosyncratic *Coelum stellatum Christianum concavum* of 1627, which had turned the constellation Pegasus into the angel Gabriel. Charles wanted to color the stars, so I read out the spectral type of each star from the *Yale Bright Star Catalogue*, and he dutifully used his Magic Markers to give each one its proper color, a characteristic level of sophistication for the Eames office, but entirely lost on the passing shoppers.

While in New York City I took the opportunity to see yet another *De revolutionibus*, one I hadn't previously recorded, at the Pierpont Morgan Library. It was my hundredth first edition, what collectors call an association copy, in this case a presentation volume from Johannes Petreius, the Nuremberg printer.

The books I had seen fell into four categories. There were a handful of

The angel Gabriel from Julius Schiller's Christian constellations of 1627, a street-side mural in the Eames Copernican exhibition at IBM headquarters in New York, December 1972.

three-star copies—on a Michelin Guide system, "worth the trip." These included the Reinhold copy in Edinburgh, the book that launched the census; Michael Maestlin's fabulously well annotated copy in Schaffhausen, Switzerland; and Harrison Horblit's presentation copy from Rheticus, the ultimate association copy since Copernicus did not himself live long enough to autograph a copy. The two-star copies—"worth a detour"—included one in Copenhagen originally owned and annotated by Matthias Stoy, one of Rheticus' students at Wittenberg who later became professor of mathematics at Königsberg. Like Horblit's three-star copy, where Rheticus had canceled Osiander's anonymous introduction "To the Reader" with a red crayon, Stoy's copy had the same red cross-out, strongly suggesting that the copy had come directly from Copernicus' only disciple. Another two-star copy was in Toronto, probably the one originally belonging to Philips Lansbergen, a seventeenth-century Dutch astronomer and table calculator. It had an interesting piece of misinformation inscribed at the end of the Osiander introduction to the effect that the Parisian scholar Petrus Ramus thought that Rheticus had written that introduction!

The one-star copies—what the Michelin Guide would merely call "interesting"—included the Morgan Library copy and one in Leningrad whose anonymous annotations contained a list of biblical verses that seemed to stand against the mobility of the Earth.

The proper classification or relevance of some of these starred copies was by no means obvious when I first examined them. Often the books became important only in retrospect, when I could make other connections with them. For example, if I had figured out who the first owner was to whom Petreius had presented the Morgan Library copy, and why that first owner had carefully canceled the Osiander introduction, it could well have won another star.

And then there was the fourth category, the large number of also-rans, with only trivial annotations or none at all. Because the books had been sold as stacks of paper that the buyer sent to the binder to be finished according to his tastes and his pocketbook, each copy was different. I carefully measured the height and width of their pages, figuring that someday this physical detail might help track a stolen copy, and I recorded the names of the previous owners whenever possible. Eventually, even these unannotated copies helped to demonstrate the movement of books and showed that the second edition, published in Basel in 1566, had particularly helped supply copies to Italy and England.

THE SPRING OF 1973 found me heading to Cairo, courtesy of American grain surpluses. In the 1960s the granaries of the American Midwest were bulging with corn and wheat, brought about in part by generous farm subsidies. Congress, having got into this particular pickle, found an ingenious way out. With Public Law 480 they arranged to send the grain to needy countries—including Poland, Yugoslavia, Israel, Egypt, and India—with the corn and wheat paid for in soft currencies. In other words, the United States found ways to spend the proceeds within each country rather than demanding hard cash. In turn, Congress allocated these funds to various government agencies including the National Science Foundation and the Smithsonian Institution.

Some of the Smithsonian's PL480 funds proved very useful for my Copernicus census, because they could be used to buy air tickets from agents within those countries. I helped organize a translation program in Poland to get certain key Polish Copernican scholarship into English, and another program in Egypt to catalog the many unexplored astronomical manuscripts from the Islamic period. Both TWA and PanAm had offices in Warsaw and Cairo, and there was no problem using the tickets they provided to stop off at various points to survey copies of *De revolutionibus*.

By April of 1973, when I was heading on my annual trip to check up on the Islamic astronomy project in Cairo, my Copernican notes were extensive enough to give a pretty good idea of the way in which the book was used in the sixteenth century, sometimes as an object of intense study, and sometimes as little more than decoration on a library shelf, casually perused at best. Having by then examined more than a hundred copies, surely sufficient for statistical purposes, I was tempted to say that enough was enough. Nevertheless, Rome beckoned, because I knew there were more unexamined Copernicus imprints in that metropolis than in any other single place. Copies were to be found in the Biblioteca Nazionale, in the Accademia dei Lincei, at the Vatican, and in the Biblioteca Casanatense. The latter library, named after Cardinal Casanate, who later became head of the Inquisition that had sentenced Giordano Bruno to the stake in 1600, unexpectedly turned out to have Bruno's *De revolutionibus*, a second edition. Bruno had been sentenced as a heretic for a plethora of heterodox ideas, including the plurality of worlds, but he seemed at best rather ill informed about Copernicus' ideas. His *De revolutionibus* contained a bold signature but no evidence that he had actually read the book. In any event,

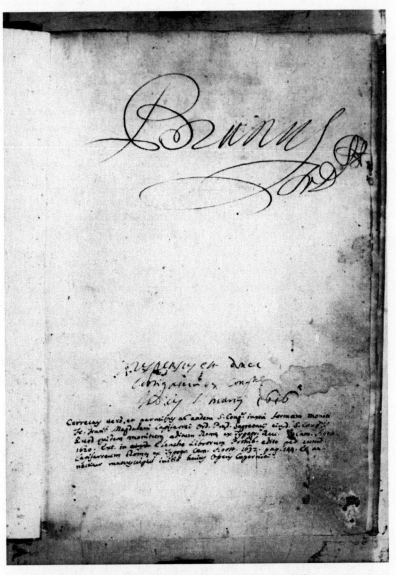

Giordano Bruno's bold signature in his De revolutionibus,
now in the Biblioteca Casanatense in Rome.

his Copernicanism was not a major factor in his conviction. Bruno's copy was a surprise, but when I arrived in Rome, the really big discovery awaited me at the Biblioteca Apostolica Vaticana.

Visitors to Rome have no trouble seeing the magnificent Vatican art collection because the Vatican Gallery abuts the edge of Vatican City, and tourists can enter to view the paintings and sculptures without needing to enter the Vatican grounds themselves. The Vatican Library is another matter, located more deeply in the territory of the Holy See. There was then (and probably still is) a visa office with full Italian-style bureaucracy. And here was an occasion when a Harvard "dazzler" letter stood me in good stead. A senior colleague had alerted me that the office of the University Marshal would prepare an official-looking document with an enormous gold seal that could help budge recalcitrant bureaucrats. Armed with my dazzler letter, I apparently passed muster. In those days gentlemen were required to wear a coat and tie. Ladies could work in the library only in the morning. In the afternoon, with no women present, men were allowed to hang their jackets over the backs of the chairs. That was one of the rules. I was asked if I wanted to see books or manuscripts.

"Books. Here are the shelf marks," I said, handing over a list of call numbers that Jerzy Dobrzycki had given me.

"But this is a manuscript," the clerk responded, pointing to one of the shelf marks on the list, Ottoboniana 1902. Puzzled, I asked for permission to see both books and manuscripts.

The Ottoboniana collection is especially interesting for Copernican studies. After the death in 1632 of King Gustavus Adolphus of Sweden, who had waged the Thirty Years' War to save northern Europe for Protestantism, with his officers helping themselves to libraries and art collections along the way, his scepter and the war loot passed to his daughter, Christina. Among other things, the twenty-two-year-old Queen Christina had hired the famous French philosopher René Descartes as a private tutor. The fifty-three-year-old Descartes, who was used to lying abed meditating till eleven every morning, was shocked by a regime that required him to get up for philosophy tutorials at 5:00 A.M. Alas, this chilly routine caused his demise, and he died in Stockholm in

1650, less than a year after his arrival. Soon thereafter Christina decided to abdicate, pack up her treasures, journey to Rome, and take up the Catholic faith. When she died in 1689, Pope Alexander VIII (the former Cardinal Ottoboni) acquired her library, whence it became part of the Vatican Library.

Knowing this piece of history, Jerzy Dobrzycki had gone to Rome for a systematic survey of the Ottoboniani, hoping to discover some unknown Copernican materials that had been confiscated by the invading Swedes and subsequently transported by Christina to Rome. Jerzy noticed that Ottoboni's collection included a copy of *De revolutionibus*, which had been classified as a manuscript volume on account of the extensive annotations bound at the back. Remembering that Copernicus had received the complete book only on his deathbed, Jerzy knew it couldn't have come from Copernicus himself, so he marched on in his survey, except that he copied out the shelf mark for me. Without that, I might never have found this copy, since it wasn't recorded in the catalog of printed books.

Inside the reading room there was another rule: only three books per day. But the library had two first editions and two second editions in the catalog of printed books, plus a rare copy of Rheticus' *Narratio prima*, and several other books I hoped to look at as well as Ottoboniana 1902. Eventually, after I had got a special dispensation from the prefect to exceed the limit, the fetchers at the circulation desk looked daggers at me. "Young whippersnapper!" they probably thought. "Who does he think he is that he can actually read six books in one day?"

One of the examples of *De revolutionibus* in the Vatican's printed books collection turned out to be quite special: a presentation copy from Copernicus' printer to the polymath Achilles Permin Gasser (who came from the same hometown as Rheticus). On the title page Gasser had penned part of a Latin poem, not a great classic but interesting:

> *By his renowned new theses, Copernicus*
> *Is believed to have put the finishing touches to this art*
> *Which Erasmus Reinhold eagerly grasped,*

> As a Thesean cord, and paved a sure path to the stars,
> And, striving to surpass the Alfonsine labors,
> Shows how great he was in the celestial art.*

This *De revolutionibus* annotated by Gasser had been part of the library of Heidelberg University, generously "given" to Graf von Tilly, the brilliant Bavarian Catholic general in the Thirty Years' War, to become a major part of the foundation of the Vatican's library of printed books.†

But the real thriller for me in the library that day was Ottoboniana 1902, and it was my turn to be dazzled. On its title page was a familiar motto: "The axiom of astronomy: celestial motion is uniform and circular or composed of uniform and circular parts." Clearly, the book had some connection with Erasmus Reinhold's copy, the book in Edinburgh that had precipitated the entire census, and which had the same motto penned on its title page. The link was confirmed by several of the extensive notes in the book itself, which matched some of Reinhold's. But there were annotations graphing the technical details of the planetary mechanisms not found in Reinhold's copy, and at the end, a wonderful series of diagrams showing at first the Copernican Sun-centered circles for the planets but then switching to Earth-centered arrangements. The transition point was specifically dated: 13 February 1578. A final geocentric diagram was labeled "the spheres of revolution accommodated to an immovable Earth from the hypotheses of Copernicus." The label seemed an oxymoron, since to us the essence of Copernicus is heliocentrism. How could the diagram have been both Copernican and geocentric?

This was extraordinarily exciting but extremely puzzling. Whose book could it possibly have been? The final diagram smacked of the geo-

---

*When in 1551 Erasmus Reinhold had issued the Copernican-based *Prutenic Tables* to surpass the *Alfonsine Tables*, he had "paved a sure path to the stars," showing how great Copernicus was as a celestial artist. Reinhold explicitly stated that the *Prutenic* (or *Prussian*) *Tables* were named in honor of both Copernicus and Duke Albrecht of Prussia.

†Heidelberg is still exercised by this perceived theft; Tilly himself was eventually defeated and fatally wounded by the Swedes under the command of Gustavus Adolphus, who had already captured Copernicus' library for Sweden.

heliocentric scheme proposed by the great Danish astronomer Tycho Brahe, not quite his final system but a logical stepping stone. In 1588 Tycho had proposed a system with the Earth at rest, with the Moon and Sun circling the Earth, but with the Sun carrying all the other planets in a retinue around it. Tycho was undoubtedly the most productive astronomical observer the world had yet seen. Night after night he had measured positions of the stars and planets, using precision instruments of his own design in the decades before the invention of the telescope. Convinced both by the reality of what he was observing and by a common-sense conviction that the Earth itself was immobile, he had sought a cosmological solution that preserved the elegant connections of the Copernican arrangement together with a solidly fixed central Earth. Published in his *De mundi aetherei recentioribus phaenomenis,** his Tychonic system bore an uncanny resemblance to the diagram at the back of Ottoboniana 1902. The Vatican diagram showed the Earth at rest in the center, with the Moon and Sun circling the Earth, but only Mercury and Venus circling the Sun. The outer planets still rode on epicycles, although a simple geometric link would have had them circling the sun, essentially switching their epicycles with the Sun's own circle. While in Ottoboniana 1902 there seemed to be some sort of embryonic generic relationship to Tycho's system, the Danish observer was already accounted for; I hadn't yet seen his copy, but it was well known to be in the Clementinum Library in Prague. The only clue to the ownership of Ottoboniana 1902 was that an early librarian had written in Latin on the title page, "embellished with autograph notes from a distinguished man." Who was that distinguished man?

Among the possible candidates that Jerry Ravetz and I had mentioned three years earlier, one stood out: Christopher Clavius, the Jesuit astronomer who had organized the Gregorian calendar reform. He had originally come from Germany, where he just might have seen Reinhold's thoroughly annotated *De revolutionibus*. In the 1581 edition of his thick

---

*"On very recent phenomena in the aethereal realm," that is, about the Great Comet of 1577.

textbook, *Commentary on the Sphere of Sacrobosco*, he had conceded only that Copernicus simply showed that Ptolemy's arrangement of the circles was not the sole possibility. If Clavius had made the 1578 notes in Ottoboniana 1902, the timing would have been just right for him to have added the remarks about alternate arrangements of circles into the revised 1581 edition of his textbook.

Leaving the Biblioteca Vaticana in a state of higher consciousness, I pondered my next move. The following day was already accounted for: A note at the hotel told me that Massimo Cimino, director of the Rome Observatory, would fetch me and show me the observatory's Museo Copernicano, which contained both a first- and second-edition Copernicus. I hoped he might also get me into the library of the Accademia dei Lincei, the famous scientific society with historical links back to Galileo's day, and which, like the observatory's collection, held both editions. Cimino arranged it perfectly, and I saw all four books the same day. One was censored according to the instructions issued by the Inquisition in 1620 (with the replacement text in a very unsteady hand); another had the same places marked but was not actually censored. And another copy had minor notes written in London in 1605, a neat demonstration of the slow reshuffling of books over time.

Cimino was helpful in another important way: He put me in touch with Father D. J. K. O'Connell, S.J., the retired director of the Vatican Observatory. I told Father O'Connell what I had found, and asked him if he could help me obtain a sample of Clavius' handwriting. He replied that the Jesuit archive was downstairs from his apartment, and that he was sure he could find something. Our paths converged in the Vatican Library reading room the next morning. Father O'Connell had in hand Xeroxes of two Clavius letters neatly bracketing the date of the annotations. We were quite excited by the prospect of what the comparison might show. Placing the letters alongside Ottoboniana 1902, we looked carefully both at individual characters and at the "ductus," the general flow of the hand. It took only five minutes to be sure that Ottoboniana 1902 had *not* been annotated by Christopher Clavius.

I left Rome in an agitated state, turning over in my mind possible can-

NICOLAI

COPERNICI TO-
RINENSIS DE REVOLVTIONI-
bus orbium cœlestium,
Libri VI.

IN QVIBVS STELLARVM ET FI-
XARVM ET ERRATICARVM MOTVS, EX VETE-
ribus atcp recentibus obseruationibus, restituit hic autor.
Præterea tabulas expeditas luculentascp addidit , ex qui-
bus eosdem motus ad quoduis tempus Mathe-
matum studiosus facillime calcu-
lare poterit.

ITEM, DE LIBRIS REVOLVTIONVM NICOLAI
Copernici Narratio prima, per M. Georgium Ioachi-
mum Rheticum ad D. Ioan. Schone-
rum scripta.

Collegij Casa-   *rei socto IESV*

Praga.   *A° 1642°.*

Ex Bibliotheca et Recognitione Tichoniana.

Cum Gratia & Priuilegio Cæs.Maiest.

BASILEAE, EX OFFICINA
HENRICPETRINA.

Title page of the "Tycho Brahe" copy of Copernicus' book in the
Clementinum in Prague.

didates, but I was stuck. The blocked-currency PL480 air tickets allowed a certain amount of creative scheduling, so I backtracked to Paris for one of the many Copernicus conferences scheduled in that quinquecentennial year. I vaguely recall chairing a session—not an easy task because several of the papers were in French and my aural French was pretty primitive. But I clearly remember my surprise at meeting one of the Czech scholars, Zdenek Horský. Normally, he had a problem getting out from behind the iron curtain, but he had ghostwritten a Copernican lecture for the president of the Czech Academy, and the trip to Paris was his reward. Horský had brought me a gift: a facsimile of the Prague *De revolutionibus* with the Tycho Brahe annotations. When I looked at the handwriting, I think my heart must have skipped a beat, for it looked suspiciously reminiscent of the hand I had just been poring over in Rome.

As soon as I got back to my Paris hotel, I checked my notes from Rome. There were too many coincidences to be accidental. Had Tycho annotated a second copy, and had I found a crucial intermediate stage in his thinking? I contacted PanAm, rebooked my flights, left Paris a day early, and headed back to Rome.

Father O'Connell accompanied me to the Vatican Library, both to smooth my way and to share the comparison between the Prague facsimile and Ottoboniana 1902. This time it took only five minutes to be thoroughly convinced that the hands matched, that Tycho had clearly annotated a second copy. The next step was to get photographs of key manuscript pages. Normally, this process could take from six weeks to six months, but Father O'Connell's presence was magic. He arranged for them to be completed in a few hours.

While we waited, Father O'Connell suggested that we go next door to the Vatican Archives to see the papers from Galileo's trial. That was an intriguing prospect, for the archival record contained not only the transcript of the infamous 1633 heresy trial but various ancillary pieces of evidence, including the famous "false injunction" supposedly issued in 1616 requiring Galileo neither to hold nor teach the Copernican system. The Galileo scholar Giorgio de Santillana had argued in his book, *The Crime of Galileo*, that the document was a forgery designed to frame

Galileo. For many years I felt that the injunction was probably a genuine document that had been prepared but never notarized because it had never actually been served while Galileo was being interviewed by Cardinal Robert Bellarmine, the leading Catholic theologian who had been instructed to warn Galileo about the dangers of holding the Copernican view. But the most modern scholarship today indicates that notarization was not required, and the injunction was quite probably actually served. In that scenario Galileo conveniently forgot about it since he had also received a letter from Cardinal Bellarmine explaining what had happened and giving a more liberal reading. That letter, introduced as evidence in the trial, is also in the Vatican file.

Alas, the archives were closed for lunch, so Father O'Connell took me above the archives to the Tower of Winds, which not many visitors see because the stairway is too narrow for traffic to go both ways. "When Queen Christina turned up at the Vatican, the very timid Pope Alexander VII installed her as far away from his own quarters as possible," O'Connell explained as we ascended the stairs. "So he put her here, underneath the old observatory. It's a very unusual observatory because it was used just with a small orifice for sunlight and a brass meridian line on the floor. That way Clavius could show Pope Gregory XIII that the Julian calendar was ten days out of synchronization with the seasons.* The walls themselves have frescoes representing the winds, and that's why it's called the Tower of the Winds."

Once we were in the frescoed room itself, Father O'Connell pointed out that the allegory of the south wind was represented by the storm on the Sea of Galilee (from the account in all the synoptic Gospels), with the opening for the beam of sunlight in the mouth of Auster, the south wind himself. O'Connell went on to say that there had long been a tradition of painting over some potentially offensive detail after Christina arrived.

---

*The meridian line was calibrated, and each noon when the solar image crossed it—far toward the south in summer, when the sun was high in the sky, and toward the north in winter—observers could determine the date. However, according to the Julian calendar, the Sun was reaching the equinox point ten days too early.

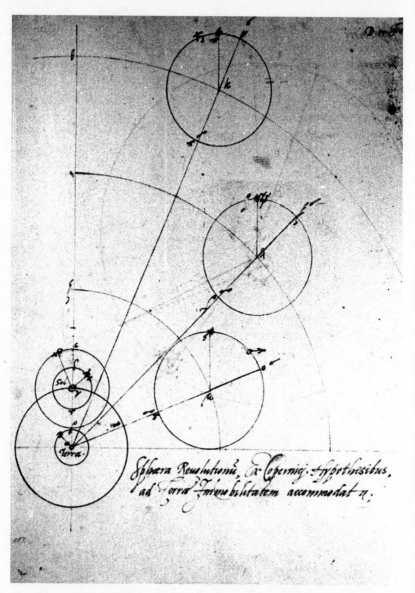

The planetary system, "accommodated to the immobility of the Earth from the hypotheses of Copernicus," from Ottoboniana 1902 in the Vatican Library.

74

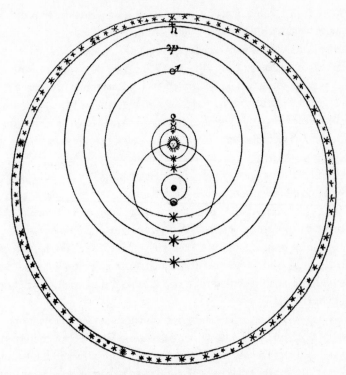

Tycho Brahe's geoheliocentric system from his De mundi aetherei
recentioribus phaenomenis *(Uraniborg, 1588).*

When restorations of the frescoes were undertaken, the overpainting
came to light: a scriptural motto under the north wind paraphrased from
Jeremiah, *Ab Aquilone pandetur omne malum* (All bad things come from
the north).

Presently the photographs were ready; I thanked O'Connell heartily
for his intervention, and carried on with the PL480 journey to Egypt.
My mind was aswirl with the unexpected serendipity of my visit to the
Vatican Library. When I had started my census, my goal had been to
find something new about Copernicus with which to celebrate the
Quinquecentennial, and here it was in spades.

BACK IN THE United States, I began to examine how the new discovery fit in with what we knew about Tycho Brahe. The Danish astronomer had published in 1588 his well-developed Tychonic system, with its fixed central Earth and with the Sun carrying the retinue of planets around it. Included was his claim that he had invented this cosmology five years earlier, that is, in 1583. But the diagrams in the Vatican *De revolutionibus* were dated 1578. However, they didn't show the completed system, and they were wrong in a crucial way. The final geocentric diagram in Ottoboniana 1902 placed the epicycles for Mars, Jupiter, and Saturn so they could glide perfectly past each other without bumping. It was as if they were made of crystalline quintessence—the heavenly "fifth element" of Aristotle—solidified and polished over the passing years of the medieval period. Unfortunately for the elegance of this arrangement, the reality of the planetary spacings required that the epicycle of Mars overlap with the sphere of the Sun. There was no question about Mars colliding with the Sun—the actual motions prevented that. But it looked unaesthetic, if not downright dangerous.

To Caspar Peucer, Erasmus Reinhold's successor as the astronomy professor in Wittenberg, Tycho wrote a revealing letter about the genesis of his system: "I was still steeped in the opinion, approved and long-accepted by almost all, that the heavens were composed of certain solid orbs that carried round the planets, and . . . I could not bring myself to allow this ridiculous interpenetration of the orbs; thus it happened that for some time this, my own discovery, was suspect to me." Finally, he realized that crystal spheres are just a figment of the imagination and not required by the Bible. Freed from the limitations of crystalline spheres, he could allow the intersection of Mars's circle with the Sun's, so that's how he depicted the Tychonic system in his 1588 publication. And Ottoboniana 1902 revealed the trail of discovery, moving step by step from a heliocentric back to a quasi-geocentric arrangement. Furthermore, because of the underlying layer of Reinholdian annotations in the book, there was a clear connection between Tycho and the Wittenberg school. "This is a tremendous scoop," I had written to Miriam from the Rome airport. "It changes several things in the accepted Tycho biography [about how and when Tycho conceived of his Tychonic system] and it dramatically confirms my hypothesis that there was an intellectual link to Tycho via Erasmus Reinhold."

I decided to use the discovery as the centerpiece of my Invited Discourse at the opening of the Extraordinary General Assembly of the International Astronomical Union in Warsaw that August of 1973. For that I would borrow a technique from Charles Eames. He had pioneered the use of multi-screen shows, such as the one in the IBM pavilion at the New York World's Fair in 1964. Afterward he had made a series of in-house three-screen slide shows, which I had seen several times at his offices in California, and I had helped him make one on Copernicus for an international symposium cosponsored in Washington by the Smithsonian and the National Academy of Sciences earlier in 1973. So I captured some slides from Eames and added many of my own, including a few new pictures of the Tycho annotations.

The Smithsonian photographer nearly balked when I told him I needed a hundred glass slide mounts for my forthcoming lecture—the actual number turned out to be 135. And my Polish hosts gulped when I told them I needed three screens and three projectionists. In the end, though, after a rehearsal, we managed to synchronize my lecture and the triple projection, which offered evocative views of Copernican Poland, of Copernicus' library, of the manuscript of his book, together with images of annotated copies of *De revolutionibus*.

IN POLAND I was shown the text of a speech to be given by the chairman of the Ministry of Health's advisory council, and was able to spare him some embarrassment. Copernicus was understandably popular in Poland that year, and it seemed every politician wanted to connect the astronomer with his own specialty. This appeared particularly promising for the Ministry of Health, because, after studying medicine in Padua, Copernicus had practiced as a doctor in his homeland. In reality, details of his medical career are difficult to come by, but the staff of the ministry's advisory council had found an article on Copernicus and the buttering of bread in the *Journal of the American Medical Association*. The article described how, during the wars against the Teutonic knights, Copernicus realized that the Polish soldiers were getting ill from contaminated bread. By placing a spread on the bread, one could tell immediately when a loaf had fallen into the dirt. So Copernicus allegedly not only carried out some hygienic research but put his findings into practice to avert an epidemic. The story involved a plausible but entirely unknown official named Adolph Buttenadt,* who popularized Copernicus' finding so effectively that the process was called Buttenadting, which was eventually shortened to *buttering*.

When I showed the article to Jerzy Dobrzycki, he roared with laughter, having spotted the fable immediately. A physician and a historian from the University of Vermont, Arthur Kunin and Samuel Hand, wishing to raise some issues about the roles of research and practice in medicine, had invented the Copernican story and the fictitious Buttenadt as their vehicle. However, the chairman of the Ministry of Health's advisory council didn't understand idiomatic English that well and thus hadn't detected the obvious spoof.

Later I mentioned the article to Professor Edward Rosen, who had made a career out of setting everyone straight on Copernicana. Rosen exploded. "That's not a hoax," he exclaimed, "that's *fraud!*" It turned out that the story had a deliciously ironic twist. By chance I met Arthur Kunin at a dinner in Vermont. Kunin told me that he had been a young student in a class taught by Rosen at City College of New York and had been so impressed by Rosen's enthusiasm for Copernicus and his attention to historical details that this probably set the stage subconsciously for their selection of Copernicus for their parable!

BEFORE THE quinquecentennial year was over, I had several more occasions to spotlight my Vatican Library discovery. One was a *Scientific American* article on Copernicus, and another was a lecture in Boulder to the largest audience I had ever addressed, approximately 1,200 persons, more people than I would have imagined to be interested in Copernicus in the entire state of Colorado. And, at the close of 1973, I repeated my three-screen show before the American Astronomical Society in Tucson.

---

*Adolf Friedrich Johann Butenandt (1903–55) won the 1939 Nobel Prize in chemistry for his work on sex hormones; can the similarity of the name in the fable be accidental?

Notwithstanding my excitement over the discovery of the annotations in Ottoboniana 1902, there remained at least two nagging questions. In Reinhold's copy in Edinburgh, he had carefully numbered and labeled the three alternative arrangements Copernicus proposed for eliminating the equant, but the technical diagrams in the Vatican copy had started in midstream "according to the second hypothesis of Copernicus." What had happened to the first? Also, there was a geographic table of longitudes and latitudes of many European cities written on a flyleaf at the beginning of the Vatican volume. Why was Wratislavia in Silesia near the top of the list, but Copenhagen and Uraniborg—two places of fundamental interest to Tycho—omitted?

The answer to one of these questions was about to come as an unexpected bombshell. The final reprise for the Copernican anniversary took place in the fall of 1974, at the annual meeting of the History of Science Society. The historians of science decided that one of the sessions would highlight the Copernican discoveries that had been made during the quinquecentennial year, and I was asked to describe the new Tycho manuscript. I drove from Cambridge, Massachusetts, to Norwalk, Connecticut, where the meeting would be hosted by the Burndy Library, an independent institution with an impressive history of science collection.* I had already inspected the Burndy's first- and second-edition *De revolutionibus* several times, most recently when I had helped dedicate a new bust of Copernicus for the library garden. Thus, instead of heading straight to Norwalk, I asked my passengers if it would be okay with them if we stopped for an hour and a half in New Haven so I could reexamine the first edition in Yale's rare book collection.

The Beinecke Library in New Haven is certainly one of the world's most beautiful. Thin polished slabs of marble let light into exhibition ar-

---

*The Burndy Library was the creation of Bern Dibner (1897–1988), a retired industrialist who had personally selected the books for his outstanding collection. In 1976 he gave the main part of the rare books to the Smithsonian Institution as a bicentennial birthday gift to his adopted nation. These books are now housed in the Dibner Library in the National Museum of American History in Washington, D.C. The remainder of the collection is found in the Burndy Library of the Dibner Institute in Cambridge, Massachusetts.

eas that girdle a central core of glass-encased book stacks. The entire architecture exudes an aura of quiet opulence (plate 8b). I already knew that the Beinecke's *De revolutionibus* was thoroughly annotated and clearly one of the most interesting copies in America. Thus, I had brought along my camera and photoflood lamps—the latter transported in a special suitcase that the Eames office had made for me—and I was installed in a small workroom. Soon the book was brought in, with its vellum binding and fragments of green silk ties that had once been able to hold the book tightly shut. As I restudied and photographed the book, I realized anew that the annotations echoed many of those in Erasmus Reinhold's copy. It had clearly been annotated in the Wittenberg circle, in two successive hands, but thus far the identity of its annotators had eluded me.

As I worked with the Beinecke copy, I slowly became more aware of its distinctive binding. Suddenly, something clicked. At the start of my survey of Copernicus' book, a particularly useful reference was a list of seventy first editions compiled in 1943 by Ernst Zinner, an eminent German historian of astronomy. Besides this compilation of locations of *De revolutionibus*, Zinner's German book on the reception of Copernicus' teachings included several other appendixes. One of them listed the sixteenth-century library of Johannes Praetorius—a student and then a teacher at Leipzig and later a professor at Erlangen—which contained two first editions of *De revolutionibus*. Zinner was a redoubtable bibliographer and loved estate lists like that, especially when most of the books had come to rest in the same place, in this case in Schweinfurt in northern Bavaria. But one of Praetorius' two copies of *De revolutionibus* was, I knew, not present in Schweinfurt. The list from 1625 (or soon thereafter) was quite explicit; it described the copy as "inn hollend. Perment unnd grün seiten Bender, cum animadversionibus Joh. Homelij et annotationibus Praetorij."* I looked carefully at the binding: Besides the Dutch vellum, here were the traces of the green silk ties, and inside there was evidence of two distinct hands. Annotations in several places cited Homelius, Rheticus' successor

---

*"In Dutch vellum and green silk ties, with the remarks of Johannes Homelius and annotations by Praetorius."

as astronomy professor at Leipzig, and these annotations were sometimes accompanied by further notes in a different handwriting, later identified as belonging to Johannes Praetorius.* I had just stumbled upon the missing book, an important link in the early dissemination of the Copernican doctrine because the annotations revealed yet another copy influenced by Erasmus Reinhold's notes.

The most amusing part in the New Haven book came at the end of a long Latin critique of a place where Copernicus was examining the length of the seasons. Unfortunately, there was a typographic error at precisely this point. Even though it didn't actually affect any of the subsequent calculations, Homelius (whose original annotations had been copied by an unknown student into the book) expressed his exasperation and finally broke into German:

> Der Himmel ist aber zum Narren worden er musz gehen wie Copernicus will. (The heavens have become a fool if they must go as Copernicus wants.)

In a state of euphoria I rounded up my passengers and headed on to the meeting at Norwalk. "I knew something had happened in there," one of them, Father Joseph Clark, later told me. "Your eyes were just dancing!"

I still have the manuscript of my talk in Norwalk, but I can't remember how much I added about my latest finding in the Beinecke Library. Mostly, I discussed my identification of Reinhold's book in Edinburgh, how it led to the Copernican census, and how I had identified Tycho Brahe's working notes in the copy in the Vatican. As was the custom at these meetings, my paper was followed by a commentator, in this case Robert Westman of the University of California in Los Angeles, a younger scholar who had come into prominence in the Copernican year. After my initial discovery of Reinhold's copy in Edinburgh, he too began scouting for other annotations in *De revolutionibus*. I knew I was in

---

*The identification was made by comparison with Praetorius' astronomical notes preserved at the University of Erlangen in Bavaria.

for some competition after Westman had sought out the annotated copy from Michael Maestlin, Kepler's teacher—a copy which I hadn't yet inspected—and had presented a widely admired paper that included a discussion of Maestlin's annotations. So there was some obvious rivalry between us, and though by the fall of 1974 I had located more than two hundred copies of the first edition, I had by no means seen them all. So I didn't know quite what to expect that Sunday morning in Norwalk.

Westman started out innocuously enough, commending my discovery and raising a few polite criticisms of my borrowed use of the twentieth-century sociological term *invisible college* to describe the network of sixteenth-century copying of annotations from one book to another. Using two projectors to produce side-by-side images, he compared the Vatican handwriting with a known Tychonic document and pointed out that they were not perfectly identical. Perhaps, he suggested, one of Tycho's assistants might have made the annotations. It was a puzzle, he said, and here was another: He quoted me as saying that the comparative pattern of annotations in the Prague and Vatican copies was made "apparently with little rhyme or reason." It seemed he was leading up to something, but what? I hadn't a clue. And then he sprang his surprise.

Westman had obtained a microfilm of a well-annotated *De revolutionibus* at the University of Liège in Belgium, and lo and behold, he had found yet another copy with handwriting matching the copies in Prague and the Vatican. This was almost unbelievable. Tycho was a wealthy nobleman, so he could have afforded three copies of Copernicus' book. But why would he have wanted *three* copies?

I was stunned. I knew at once that I couldn't discuss Ottoboniana 1902 definitively without taking into account the newfound Liège copy. I had been aware of a first edition there, but it had not been my procedure to ask for a microfilm before I had actually looked at a book. Westman had stolen the march on me by writing letters to many institutions and requesting microfilms when the descriptions sounded promising. The trauma of his coup has completely blotted out my memory of who else was on the program that Sunday morning.

Half a dozen of us went out to lunch together that noon, across the highway from the Burndy Library at Old McDonald's Farm, a rather funky theme restaurant saturated with rural atmosphere, and a favorite watering hole on trips from Boston to Manhattan. Against a background of old farm implements and circus posters, Westman and I approached the situation as warily as scorpions in a bottle. Gradually, the tensions ebbed as we realized that the only rational resolution would be to join forces in describing the Tycho Brahe material. Although there would be tempestuous days in the future, it was in retrospect one of the happiest decisions I've ever made. If nothing else, I had much more fun working with a colleague than in isolation. Our research strengths were remarkably complementary, and what we eventually produced was far more interesting than either of us could have managed alone.

The Liège copy resolved a principal mystery surrounding Ottoboniana 1902. It contained notes in two different hands. The earlier layer was a remarkably exact copy of Reinhold's notes. This, then, was the source for the title page motto and all the other Reinhold material in the Vatican copy. That early layer of annotations also included a page of notes, apparently from Reinhold, but now missing from the Edinburgh copy—an analysis of the planetary circles "according to the first hypothesis of Copernicus." With that analysis already in place, it was no longer puzzling that the Vatican annotations began with the "second hypothesis."

But there was still no clue why the table of geographic coordinates in Ottoboniana 1902 listed Wratislavia so conspicuously and omitted the key Tychonic locations, Copenhagen and Uraniborg. That would come back to haunt us.

Chapter 6

THE MOMENT OF TRUTH

JERZY DOBRZYCKI'S dry sense of humor always enlivened my visits to Warsaw. He had a stoic attitude of realism about coping with conditions there, which meant long hours of standing in lines. Even that sometimes had its ironically funny moments. The Staszy Palace, home of the History of Science Institute, was a hotbed for the Solidarity movement, and there came a time when the entire building was placed under siege by the military and its occupants were arrested. Jerzy and his colleague Paul Czartoryski, who edited the series of Copernican studies and the *Complete Works*, were so busy at work that they didn't want to leave immediately. "You must come," said the departmental secretary, Olga. "This is what Solidarity means; we're all in this together." So at last Dobrzycki and Czartoryski went down to the front portico of the palace, only to be faced with a long line of workers waiting to be arrested and a large crowd of curious onlookers. As they paused, wondering what to do, a policeman came by and suggested they get lost. And so they did, melting seamlessly into the crowd.

But in August of 1973 those events were still some years in the future, and Poland bubbled over with a well-developed Copernican euphoria. The quinquecentennial celebrations were in full swing, colorful Copernican posters were everywhere, wooden plates with the astronomer's portrait were found in every kiosk, and international visitors were coming to trace the footsteps of Copernicus. Included were scientists from around the globe, assembling in Warsaw to pay homage to their illustrious forebear.

Of course I had told Dobrzycki what I had found at the Vatican Library before giving my IAU invited discourse, but he was still chagrined that he had passed over the volume when he had first seen it in Rome. Nevertheless, he continued to back my census with full enthusiasm. We went together to his hometown of Poznan to see the six *De revolutionibus* copies in the libraries there, and then to Wrocław, where there were two important collections, one at the university and the other at the Ossolinski, an independent institution that had transferred from the vicinity of Kiev just after World War II. Dobrzycki showed me the printed catalog of the Ossolinski books, and I noticed that for every entry there was an ordered list of all the provenances, that is, all previous ownerships and locations. This arrangement seemed like an ideal approach for the type of extensive survey that I had undertaken, although it also meant that much of the information I had collected wasn't fully adequate.

Dobrzycki agreed on both counts. Clearly, the census was undergoing a significant metamorphosis. It would not simply be a matter of checking whether the copies were annotated. Not only would I redouble my efforts to inspect nearly *all* of the first and second editions, but I would need to be more systematic in collecting background information on each copy so that where possible I could track who had owned it over the centuries. So, a few years into the survey, I realized that I would have to go back to many of the libraries to make sure that my notes were complete enough. This meant returning to Cambridge, Oxford, and London to ensure that I had the provenances of several dozen books.

I'm often asked how I found where the copies are. If you live in Europe or North America and want to see just one copy, that's easy.* There are several primary reference resources to track one down. Today the OCLC (On-line Computer Library Center) or the KVK (Karlsruher Virtueller Katalog) gives immediate on-line assistance. When I began, the *NUC (National Union Catalog)*, a magnificent reference work of 400 volumes, provided a starting point, though one always has to be on guard against several

---

*But if you live in Cape Town, the nearest first edition is about 6,000 miles north, in Naples. From Buenos Aires it is about the same distance to Guadalajara, Mexico. From Sydney the nearest first edition is in Manila, nearly 5,000 miles north. From Delhi it is almost an equidistant 4,000 miles to first editions in Moscow, Manila, or Hiroshima.

twentieth-century facsimiles being mistaken for the original.* But the *NUC* listed only about one-third of the forty-some copies of the first edition in North America and of the slightly greater number of second editions, so if you want to see *all* the copies, you have to use additional tactics. One of the principal ways to find other copies is to write letters of inquiry to old or large libraries. Another is to ask specialist book dealers. At any one time there are worldwide scarcely two dozen firms with serious inventories of rare science books, and I contacted them all. One of the most helpful dealers was Jake Zeitlin, a major figure in the cultural life of southern California. Zeitlin was very forthcoming in helping me locate books in private collections. He had, for example, sold a copy to a distinguished medical doctor who had subsequently grown senile, but he was determined that I should see it. He took me to the doctor's mansion, and we waited till the nurse had wheeled the collector out to the swimming pool before we made our surreptitious entry into the library. The copy had no annotations, so it was a quick job to measure it (one of the largest copies in existence because the binder had trimmed very little of the rough edge of the paper) and to document its binding without its owner having the slightest inkling of the intellectual burglary being pulled off inside.

I was particularly keen to see the Liège copy that Robert Westman had surprised me with at the 1974 History of Science Society meeting, and I got my chance soon after that session in Norwalk, on my way back from one of my periodic inspection visits to Cairo. I stopped off in Brussels and drove to Liège, bringing along my suitcase of photoflood lamps in order to make some slides of the book.† To see for myself the complexities of these annotations was an exhilarating experience. The Liège volume had two layers of handwriting: an early hand that had simply copied Reinhold's anno-

---

*Facsimile editions of the 1543 first edition were published in 1928 (Paris), 1943 (Amsterdam and Turin), 1960 (Leipzig), and 1966 (Brussels), and a CD Rom in 1999 (Los Altos). A facsimile was also made of the 1566 second edition in 1972 (Prague).

†One of my most distinct memories of this trip is the fact that Liège was the only place where I have ever found 165 volts in the electric lines. I traveled with photoflood bulbs for the standard European 220 volts, so I was temporarily stymied, but presently the astronomers produced a transformer and I was in business.

tations, and an overlay of notes by Tycho Brahe. But the early hand included some material that was very likely from Reinhold yet was no longer found in the Edinburgh *De revolutionibus*, and it was precisely the material on these pages that had triggered the entire sequence of planetary orbits in the well-annotated Ottoboniana 1902 in the Vatican. The plot was thickening, but it would be some time before Westman and I could sit down together to come to terms with the Tychonic material.

Meanwhile, I worked assiduously to examine other, far-flung copies, taking advantage of European conferences or my annual trips to Cairo to visit libraries throughout the United Kingdom and the Continent. In those days it was difficult to see the forest for the trees; the fascinating connections between many of the volumes became apparent solely in retrospect. For example, I realized only gradually what an extraordinary collection of copies survives in Scotland. In Edinburgh alone there are half a dozen copies. Besides the seminal exemplar annotated by Erasmus Reinhold, there is one owned by Adam Smith, the economist who wrote *The Wealth of Nations* and who alluded to *De revolutionibus* in his essay on the history of astronomy. The copy won three stars in the *Census*, not because of Smith's ownership but on account of an earlier and still-unidentified owner from the Wittenberg circle, who had written, "Anyone can rightly wonder how from such absurd hypotheses of Copernicus, which conflict with universal agreement and reason, such an accurate calculation can be produced, and why he did not undertake the correcting of the Ptolemaic hypotheses, which agree with Sacred Scripture and experience, rather than producing such a paradox." The description of this copy, with its anonymous defense of Copernicus, takes five pages in the *Census*, and I would gladly offer a champagne dinner to anyone who can establish who the annotator was.

There was yet another well-annotated copy in Edinburgh, owned by the physician John Craig. This book would later play a key role in untangling the Tychonic annotations and would unexpectedly confirm a hoary rumor about the invention of logarithms. In Glasgow I found three first editions, but not until the final days of assembling the camera-ready copy for my *Census*, decades later, would I realize that one of them had a particularly remarkable provenance. In Aberdeen the librarians simply locked me in the rare book stacks, so I had a marvelous opportunity to see the range of six-teenth-century astronomy books brought back to Scotland by Duncan Liddel, who had taught astronomy in various universities on the Continent in the 1580s and 1590s. His second-edition *De revolutionibus* was clearly a three-star copy, for it contained on interleaved pages the third known six-teenth-century copy of Copernicus' *Commentariolus*, the one identified by Jerzy Dobrzycki; but more than that, the marginalia in this copy would eventually connect with the Tycho story.

At St. Andrews (perhaps more famous for its golf course than for its university) there was a first edition that had once belonged to a "German nation"—one of the Renaissance university student guilds or unions for German-speaking students. The entire slate of officers was duly inscribed on the flyleaf. Unlike many of the copies in the British libraries, this one had not been acquired until the nineteenth century, as an antiquarian item, so its earlier provenance was quite obscure.*

IN CONTRAST to the richly annotated copies of Scotland, those in the French provinces seemed like a barren intellectual wilderness superim-posed on an inviting touristic landscape. Yet ultimately, the very sparsity of annotations, and in particular, the lack of censorship demanded by the Roman Inquisition in 1620, yielded one of the unexpected and most in-teresting insights of the project.

In the summer of 1976 the International Astronomical Union met at Grenoble in southern France, giving an occasion for one of several cam-paigns to visit a number of libraries scattered around the country. Two in-vestigations had enabled me to locate these copies. The first I undertook at the Bibliothèque Nationale in Paris. Its reference section contains printed

---

*Quite by chance I discovered in which university town the "German nation" of the St. Andrews copy was actually located. While in Padua, as I was looking at an exhibition marking the four-hundredth an-niversary of Galileo's professorship there (in 1592), I noticed in a display case the same roster of names in one of the exhibited books; a further investigation disclosed that they were all among the foreign stu-dents at Padua in the early 1600s.

catalogs of numerous provincial libraries; these volumes seem unique to France and are a monument to the Gallic penchant for systematization. Arranged by a rigorous subject classification, they could be swiftly checked for Copernicus' book, and I surveyed dozens of them with a comparatively modest investment of time. The second net was cast by René Taton, dean of the French historians of science, and Maylis Cazenave; they advertised in a journal read by French librarians, and they wrote letters to likely libraries. Between these two surveys, we located seventeen first editions and sixteen second editions in the French provinces, which rather astonished most of my Parisian friends, who like to think that a majority of the important treasures have long since been centralized in the capital.* In any event, these lists served as a good guide for scholarly tourism in the French outback.

Following the astronomical sessions in Grenoble, Jerzy Dobrzycki (who by then had succeeded me as chairman of the IAU's History of Astronomy Commission) went with Miriam and me to nearby Vienne. We admired the Temple of Augustus, an antiquity from Roman times and Vienne's two-star wonder in the *Guide Michelin*, but our goal was the Bibliothèque Municipal, which turned out to be a matchbox-sized library facing a nearby square. Its first edition had only minor annotations, but it bore a splendid signature on the title page: Pontus de Tyard, a key French author of the sixteenth century. Pontus belonged to the Pléiades, a coterie of seven young men who exercised a formative influence on their native language; their purpose was to encourage the writing of French as against Latin, and to enrich the literary language with appropriations from classical Latin and Greek. Pontus wrote sonnets as well as a wide-ranging discourse on the parts and nature of the world entitled *L'Univers* (1557), which gave several laudatory references to Copernicus, and he even authored a technical and now extremely rare *Ephemerides octavae sphaerae*. With his copy of *De revolutionibus* in hand in the library, I coveted a slide of the title page with his autograph, and looked around for a window with

---

*There are a dozen first editions in Paris, and a baker's dozen of the second.

enough light for my Nikon. In the absence of a suitable window, the librarian suggested that I simply take the book out into the square in front of the library. I've always regretted that I didn't have Miriam snap a picture of Jerzy and me examining the book on the front steps of the Bibliothèque Municipal in Vienne.

That afternoon we drove south to Lyons; Miriam and I soon found the public library, a handsome building in a brand-new shopping mall, which contained two nonmemorable first editions. Rush-hour traffic impeded our departure, so we made slow headway to our next destination, Clermont-Ferrand. The next morning we found the library in Clermont-Ferrand by 11 A.M.—leaving very little time to inspect the *De revolutionibus* before its 11:30 closing. There was a bit of a flap finding the book, as it turned out to have already been fetched and placed on reserve for me. Since it wasn't annotated, recording its size, binding, and condition for the census was easy. Like most of the books, it did include early ownership inscriptions, so I could see that it had been in a Carmelite convent in Clermont already in the seventeenth century. Clermont—Blaise Pascal's hometown—lies on the slope of Puy de Dôme, the extinct volcanic peak where in 1648 Pascal had carried a primitive barometer to demonstrate how the weight of the Earth's atmosphere balanced the mercury in the tube of the barometer. We drove to the top, only to find it wreathed in thick fog; not until we were leaving town did the clouds lift long enough for us to glimpse its summit.

That evening we made it to Bourges, got a hotel with bath and breakfast for fourteen dollars, and prepared for our visit to its library the next day, Saturday. Again, there were no annotations, so we hastened on to Troyes, where two more lightly annotated copies were available—wondrous to report—on a Saturday afternoon.

This 1976 field trip was typical of many that crisscrossed Europe and which enabled me to inspect personally nearly every locatable copy. The successive European campaigns yielded a kaleidoscopic view of *De revolutionibus*, some memorable, some hopelessly nonmemorable but in memorable libraries, and some merely faded blurs of memories. The journey from Grenoble to Vienne and north to Paris, for example, demonstrated that libraries throughout the land had acquired Copernicus' book early on—*De revolutionibus* was clearly perceived as a work that any serious collection required. The copies in Lyons, Clermont-Ferrand, and Bourges had all been owned by clerical libraries, and not one had seen the censor's hand, even in a Catholic country. Thus, despite the absence of annotations, a significant pattern was beginning to emerge.

THE SUMMER OF 1977 brought a change of pace to the Copernican research. Six years had elapsed since my Copernicus quest had begun, and another sabbatical year in Europe was about to begin. For the first semester I was a visiting fellow at St. Edmund's House in Cambridge. Once a virtual monastery for Catholic male students, it had become quite ecumenical. Miriam, our youngest son, Peter, and I settled down in a maisonette on St. Edmund's grounds. Not many miles away, Robert Westman had taken up residence as he, too, was taking a sabbatical year and had likewise chosen Cambridge for its venue. Although we were working on several independent projects, we periodically got together at the Observatories, where I had both an office and the loan of a microfilm reader. Meanwhile, I continued my pursuit of still more copies of *De revolutionibus*.

It was well known in history of science circles that one of the most important collections of scientific manuscripts from the time of Isaac Newton was locked away in Shirburn Castle, the Earl of Macclesfield's residence near Oxford. In the eighteenth century the 3rd Earl had been a devotee of astronomy, and one of the principal backers of calendar reform in the House of Lords when England and its colonies (including America) finally accepted the Gregorian calendar in 1752. His assistant, William Jones, had been a tireless collector of books and manuscripts, which had gone to the Earl after Jones' death in 1749. In 1896 the grandson of the 6th Earl had inherited the title while a mere child, and, as the 7th Earl, had lived to an old age. This posed a formidable problem to historians of science throughout most of the twentieth century because the 7th Lord Macclesfield had fended off historians. It was only by a special concession that the Royal Society, Britain's leading scientific organization, had been allowed to copy the rich repository of Isaac Newton letters in the collection in order to complete its magisterial seven-volume edition of Newton's correspondence.*

I suspected early on that Shirburn Castle was a likely site for a copy of Copernicus' book, but I assumed access would be next to impossible. One day a colleague remarked that the newly titled 8th Earl of Macclesfield might prove a little more amenable to showing his treasures. After the 7th Earl had died in 1975, the Inland Revenue office had struck an agreement as part of the death duties settlement to the effect that the collection was supposed to become more easily available to qualified scholars, but this agreement was apparently a well-kept secret. My colleague suggested that if Lord Macclesfield had a *De revolutionibus*, he might be willing to deposit it in the Oxfordshire Public Records Office for a few days so that I could examine it. Armed with this fresh intelligence, I asked my office in the States to send a letter of inquiry. By return post came the answer: Both the first and second editions were in the collection, and I was instructed to phone the earl. It took a few days for me to gather the courage to place the call. When I asked him if he would deposit the volumes in the Oxfordshire Public Records Office for me to inspect, he surprised me by inviting Miriam and me to Shirburn. At his suggestion we settled for the following Thursday.

As castles go, Shirburn is not pretentious, but it wins high honors for its bucolic charm and for the moat that completely surrounds it. We entered on the side via a small permanent bridge. "How nice of you to come on Thursday," Lady Macclesfield said. "This is the only day we have a cook, and we can invite you for lunch."

We were ushered into a drawing room decorated by two large paintings of horses on each of the four walls. I straightaway recognized the eight Shirburn canvases as the work of George Stubbs, a fashionable English painter of the eighteenth century. Lady Macclesfield led Miriam away for a tour of the castle, which included cranking down the main drawbridge, while Lord Macclesfield and I turned to the two

---

*In 2001 Cambridge University Library acquired the collection of letters for £6.37 million.

copies of *De revolutionibus*, which he had brought down from the library for me.*

Of the two, the first edition turned out to contain some unexpected annotations, quite unconnected with Copernicus' text. The copy had formerly belonged to John Greaves, an early-seventeenth-century professor of astronomy at Oxford and a pioneering scholar in Islamic scientific texts. Greaves apparently acquired his exemplar in Italy while en route to the Holy Land, and its endpapers became a convenient notebook when he visited Aleppo in Syria. There he had the opportunity to copy some remarks from an earlier traveler who had gone overland to the Mogul court in India, and who had recorded in his diary that he had "observed two unicorns with his own eyes." Lord Macclesfield had never had occasion to notice the inscription before and was quite amused when I read it to him. Only much later, after my *Census* was published, did I learn that the traveler who had seen the unicorns in Lahore was Thomas Coryate, a notable seventeenth-century English traveler whose published diary included a fanciful woodblock of the mythical beast.

Greaves was one of the few scholars to use a pencil rather than a pen when he annotated the text itself, particularly the mathematical sections. On the otherwise blank leaves inside the front cover he had included numerous small diagrams and rules for spherical trigonometry, as well as correlations of chronologies, partly in Arabic and Persian. In contrast, the second edition, which had been owned by an early English mathematician appropriately named Euclid Speidell, was unannotated apart from the owner's name. That turned out to be fairly typical: Of the fifteen copies of

---

*Over lunch the Macclesfields explained that Shirburn was one of only three or four buildings in England that still retained a working moat on all four sides, and that cinema companies frequently asked to rent the castle. But Lord and Lady Macclesfield considered this to be too much of a nuisance since the movie moguls would wish to pull down the anachronistic rain gutters. Lady Macclesfield also remarked that they had occasionally thought about opening the castle to the public on Sunday afternoons, for besides the paintings it included many historical curiosities such as Queen Elizabeth I's riding gloves. Unfortunately, they noted, the flow pattern was not conducive to large groups, so nothing ever came of that. As far as I know, Miriam is the only person in history of science circles who has had the tour.

The unicorn from Thomas Coryate's travels to India in 1616. That the woodblock is trimmed suggests it was recycled from an earlier book.

the first or second edition in Oxford, nine were essentially unannotated apart from ownership marks.

BOB WESTMAN had joined me at the Cambridge Observatories for one of our joint research sessions. As a warm-up exercise, we threaded into the viewer a microfilm of *De revolutionibus* that I had ordered from the University Library in Wrocław, Poland. The neat and extensive annotations grabbed our attention at once. As an inscription on the title page revealed, they had been made around 1600 by one Valentin von Sebisch, a city councillor at Liegnitz in Poland, but a nobody as far as astronomy was concerned. This fact in itself puzzled us, but there was another problem. The annotations were just too neat, without the cross-outs or interlinear interpolations so characteristic of working notes, and they were too perceptive. Sebisch, for example, pointed out that one of Copernicus' purported observations exactly matched the calculated position from the venerable *Alfonsine Tables*. But then something really took us by surprise: We suddenly realized that Sebisch's marginal drawing and part of the annotation on folio 113 agreed closely with Tycho's notes on the same page of his Prague copy. Our discovery explained at once why Sebisch had seemed too clever for a nobody and too error-free in drafting his notes. His annotations were surely just a straight copy from Tycho Brahe.

We had conquered one mystery, but a bigger dragon rose out of this solution. Sebisch's notes matched Tycho's copies here and there; they were obviously from the same family, but just as clearly they seemed to stem from a Tychonic copy we had not identified. Between the two of us, we had canvassed the major universities and libraries well enough to know that the chances of finding a fourth copy from Tycho were pretty low—and yet we found it within six weeks!

My earlier search of the auction records had left a couple of loose ends, one old and one new. First, the new one: Several recently auctioned copies had been knocked down* to a buyer named Umiastowski, whose English address was unknown to me. My Polish colleagues identified him as the colonel who, at the time of the German invasion of Poland in 1939, had gone on the radio to urge every able-bodied man to the front, and had thereby created the most monumental traffic jam yet known in Warsaw. The old loose end was that in 1912 and 1913 Christie's in London had auctioned first editions, but I had been able to trace only one of them. So I asked Christie's if I could look at their old records in search of clues to help identify the unlocated book.

In the workroom at Christie's I soon found the answer to the older problem. The 1912 and 1913 copies were one and the same. For the first sale, the description "with marginal notes and annotations" had been deemed inadequate by the London bookseller Bernard Quaritch (who had won the *De revolutionibus* with a bid of twenty-two pounds), so the book was returned because the auction catalog hadn't indicated that folios 38

---

*When the auctioneer taps his or her hammer to indicate that the bidding is over, the item is said to have been knocked down.

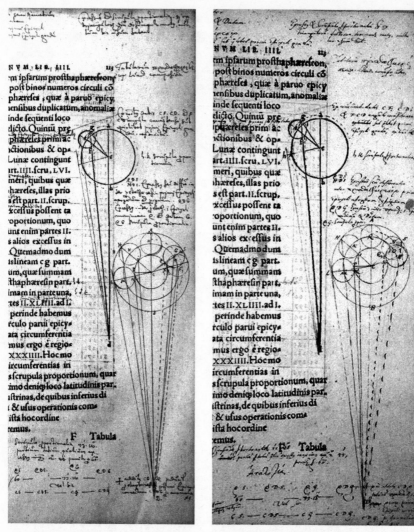

Valentin Sebisch's marginal annotations in De revolutionibus *(left), neatly copied from the original set made by Paul Wittich (right). Both copies are now in the Wrocław University library.*

and 39 were in an old calligraphic facsimile, and not the original printed leaves. Christie's corrected the description by specifically mentioning that two leaves were supplied in manuscript, and put it on the block once more, four months later. Quaritch again purchased it, but this time for much less. My subsequent researches revealed that Quaritch's still had the book after World War I, and in 1925 they were offering it for twenty-five pounds. Six years later, despite the stock market crash, the London firm of Henry Sotheran had it, priced at seventy-five pounds.*

As I was zeroing in on the identification of the ghost copy, the telephone rang. A Christie's staff member held a brief conversation, hung up, and turned to a colleague to ask, "Do you know anyone named Umiastowski?"

His coworker merely shrugged, but I leapt to attention. When I discovered that Umiastowski was in the waiting room, since the person he wanted to see wasn't in yet, I rushed downstairs and quickly guessed who was the expatriate Polish colonel. I introduced myself and launched into a conversation about his famous countryman. Umiastowski warmed to the subject and expansively claimed to have four copies of Copernicus' book. Before the conversation was over, I had got his address and an invitation to come and see his books in Dulwich in the near future.

By mid-December 1977 I had made the arrangements to visit the eighty-six-year-old colonel. Roman Umiastowski began by showing me his first edition—a "complete" copy, he proudly announced. I was quite

---

*The copy could well have accidentally come incomplete from the Frankfurt Book Fair, and rather than return it, an original owner may have made the manuscript facsimile of the missing signature. The fact that this copy was imperfect accounts for its later wandering. It was much less expensive than a perfect copy and thus affordable by an English mathematics professor who taught for over a decade at Iowa State University. Apparently, he sold it around 1937 when he returned to England, and it was bought by Yale University. Eventually, a Yale benefactor offered to exchange this one for a perfect copy (the most important first edition in America, described in the previous chapter), and in 1971 this benefactor gave the imperfect copy to Stanford University. When Stanford subsequently acquired a perfect copy, I realized that this one might be released as a duplicate. I knew that the staff in charge of special collections at San Diego State University greatly coveted a copy, but the library could not afford the astronomical price to which a perfect copy had ascended, so I suggested that they negotiate with the Stanford library. In 1991 the library at San Diego State University purchased the imperfect copy as its millionth book acquisition.

puzzled and not a little suspicious since the book had been auctioned rather cheaply a few years earlier as imperfect, lacking eight leaves. As I paged through it, I was thrilled to recognize that its marginalia matched an extensive set I had seen in Toronto. I was slowly coming to realize that major annotations seldom come singly; before my eyes a new family of annotations was emerging.

I turned another page and was stopped cold in my tracks. The marginal handwriting totally changed, but what arrested my attention was the fact that this second hand was Tycho's! By that time I was so familiar with his distinctive writing that I could be quite sure, yet for a moment nothing made any sense. I paged on, and eight leaves later the first handwriting reappeared. Very quickly I realized that Umiastowski had inserted eight Tychonic leaves to make his first edition complete; in fact, the leaves were not from a first edition at all but from a second.

I didn't have to look far to see the source of the eight Tychonic leaves. Stacked on the far side of the table were the folios from a broken *De revolutionibus*. Besides the broken Tychonic copy, there were two more second editions as well as the first. Umiastowski's great joy was to make books complete, and having bought several imperfect copies, he was in the process of disassembling one of them to make other incomplete copies complete. The problem was, Umiastowski was cutting up what was obviously the most important second edition in private hands in order to repair the others. It was theater of the absurd. I didn't know whether to laugh or to cry.

As diplomatically as I could, I explained the importance of the annotations to the colonel. Umiastowski looked very sad and allowed that he had just brought his "complete" copy back from the bindery the previous week. Before I left, he had agreed to reassemble the Tychonic copy and to allow me to return to make a microfilm of it. I could hardly wait to get back to Cambridge to recount my adventure to Bob Westman. We were both almost incredulous at the idea that I had so promptly found the original notes from which Sebisch had copied. By January I had made my microfilm, and we began to pore over the annotations. But Bob was increasingly troubled, and he recalled his original doubts about the handwriting.

"There's something wrong," he asserted. "Why would Tycho Brahe, who was so industriously building instruments on his island of Hven, and so busy making observations, take time to annotate four separate copies of *De revolutionibus*? At the very least he must have used an amanuensis."

I objected, saying that if the copies were all alike, of course a secretary could have copied the notes, but these copies were each different. And you don't tell your secretary to write in the margin, "I don't understand this theorem." Such a note had to be written by the astronomer himself. But I agreed with Bob that we had to check the handwriting to make sure it really was Tycho's. We knew that the great bulk of Tychonic manuscripts were in the manuscript collection of the Austrian National Library in Vienna. I remembered I had some Vienna photocopies back in my office in the States, but we were in a hurry. Then I recalled that sometime in the nineteenth century one of the Czech scholars had produced a facsimile of a trigonometry manuscript in Tycho's hand. The next day, I found it in the Cambridge University Library. Titled *Triangulorum planorum et sphaericorum praxis*, it was printed in Prague in 1886.

The trigonometric manual turned out to be quite problematic. The agreement between the handwriting in the old Prague facsimile and in the annotated copies of *De revolutionibus* was not very convincing and left a cloud of ambiguity over the whole project. Bob was convinced there was something fishy about the Prague document, and pretty soon he retrieved more information about it in the multivolume *Tychonis Brahe Opera Omnia*, the modern edition prepared early in the twentieth century by the astronomical historian J. L. E. Dreyer. Dreyer was highly critical of the claims made for the Prague trigonometry manual. It couldn't be in Tycho's hand, he declared; Tycho never made a two-stroke capital *M* like those in the *Triangulorum planorum* and also, disconcertingly, like those in the annotations in our four "Tycho Brahe" Copernicus volumes.

Had we been deceiving ourselves for four years? Was it possible that Tycho wasn't, after all, the author of those interesting planetary diagrams in the Vatican?

I put in a phone call to my office in the United States, requesting that my photocopies of some of the Viennese Tycho materials be rushed to me at once. About a week later the sheets arrived. Bob and I eagerly compared them with the microfilms of the Copernican annotations. It did not take long to realize that we were in trouble. The handwritings did not match at all.

The moment of truth had arrived. We had been chasing the wrong scent. Tycho Brahe could not possibly have annotated those four copies of *De revolutionibus*. As Kepler said when one of his ideas collapsed, "The hypothesis has gone up in smoke."

Chapter 7

THE WITTICH CONNECTION

ON SATURDAY, 29 October 1580, the noble-born, imperious, and eccentric Tycho Brahe inscribed the title page of one of the most lavish and spectacular books printed in the sixteenth century (plate 6). Peter Apian's *Astronomicum Caesareum* was truly astronomy for an emperor's eyes. Dedicated to Charles V of Spain, it had won for its author, an astronomy professor at the university in Ingolstadt, the right to appoint poets laureate and to pronounce legitimate children born out of wedlock. A giant folio with brightly hand-colored pages, not only was it a tour de force of scientific printing, but its numerous sets of volvelles—layers of rotating paper disks—could be used to calculate planetary positions and configurations. Tycho later admitted it had cost him twenty florins, which by today's currency would be roughly $4,000.*

The inscribed *Astronomicum Caesareum* was a princely gift for a gifted visitor, one Paul Wittich, a peripatetic mathematical astronomer from central Europe. Wittich had been on Tycho's island fiefdom of Hven for about six weeks. He had admired Tycho's quadrant and sextant, had examined the clever scales on the instruments that allowed Tycho to read the angle to a minute of arc, and had heard about Tycho's plans to build a new and larger quadrant, to be affixed to a main inner wall of the Uraniborg castle. Wittich had brought along some ingenious mathematical tricks for converting stellar coordinates from one system to

---

*Copies today sell in the vicinity of half a million dollars.

another, much admired by Tycho, and he had some stimulating ideas about the technical details of planetary cosmology. Clearly, Tycho greatly appreciated his visit and hoped that Wittich would return. The big book was part of the strategy. "To Paul Wittich of Wratislavia, friend and fellow lover of mathematics," Tycho boldly wrote in Latin.

I pondered the relationship between the well-known Brahe and the somewhat shadowy Wittich as I drove from Cambridge to Oxford one morning in late January 1978. I was scheduled to give the astronomy colloquium on my latest Copernican research. The Tychonic bubble had burst a few days earlier, and I was still trying to reassemble the pieces. Wittich seemed part of the puzzle. I had found a fascinating set of annotations in a second-edition *De revolutionibus* in Wrocław when Jerzy Dobrzycki and I were there in 1974; they included marginalia copied from both the annotated Liège volume and from Kepler's teacher, Michael Maestlin. There were some mathematical problems that used 52° for the latitude, close to that of Landgrave Wilhelm of Hesse's observatory in Kassel. Today Hesse, lying in west-central Germany, is the eighth largest of the sixteen German states, but Wilhelm's powerful father, Philip the Magnanimous, had divided his territory among his four sons, and Count Wilhelm ruled only the northernmost principality, Hesse-Kassel. Wilhelm owes his lasting fame to his research in astronomy, and his observatory was second only to Tycho's. The rolls of astronomers there included Paul Wittich, who seemed an obvious candidate to be the annotator of the Wrocław book, but my attempts to locate a sample of his handwriting had failed. Still, in my mind's eye, I had assigned the book to Wittich.

I had found two more samples of what I thought to be the "Tychonic" hand besides those in the quartet of Copernicus books. One was in a rare copy of Tycho's book on the new star of 1572, which I had taken as Tycho's working notes. The other was in that splendid presentation copy of the Apian *Astronomicum Caesareum*, now preserved at the University of Chicago Library. I supposed that Tycho had made a few marginal notes before magnanimously handing over the inscribed volume to Wittich. Somehow the simple logic that Wittich himself wrote in the book after he got it eluded me on the drive to Oxford.

Tycho Brahe, from Albert Curtz's
Historia coelestis *(Augsburg, 1666).*

My musings turned to two sets of annotations derived from the four pseudo-Tycho *De revolutionibus* copies, which I had seen in the summer of 1975. Both had been made by Scots working on the Continent and brought back to their homeland. Duncan Liddel of Aberdeen taught some years in Rostock, and in 1587 he paid a week's visit to Tycho's Uraniborg castle. John Craig of Edinburgh had been dean at Frankfurt an der Oder for several years but later returned home and became personal physician to James VI (later to become James I of England). The king had paid a state visit to Uraniborg in 1590, and quite possibly Craig was part of the retinue. Craig's books passed to the king's secretary, who eventually donated them to the Edinburgh University Library.

By now I was in the outskirts of Oxford, and, still puzzled, I forgot about Tycho, Wittich, and the Scots, and concentrated on the traffic. I broke the news to my Oxford audience that Tycho was not, after all, the draftsman of the wonderful diagrams in the Vatican Copernicus.

The next afternoon, as I drove back to Cambridge, all at once I remembered an essential clue. Both Liddel and Craig had been tutored by the mysterious Paul Wittich. And something else clicked. In the Liège copy, the one whose annotations Bob Westman had first noticed, there was a marginal note in the first person: "the mean motion most exquisitely determined by me." Craig's copy in Edinburgh read somewhat differently. It said something about "Witt," which I had taken as an abbreviation for Wittenberg. But what if instead it stood for Wittich? Did the note say that Wittich himself had determined "the mean motion most exquisitely"? I could scarcely wait to examine the microfilms, but it was hard to drive much faster on the winding route from Oxford.

As soon as I got back to Cambridge, I confirmed the reading in Craig's copy. How could I have been so dense not to have noticed it before? I telephoned Bob Westman. "I know who annotated the Vatican copy. It was Paul Wittich." Bob was skeptical but soon came around to the new view. Our task was then clear: to find out everything possible about the elusive Paul Wittich. To begin with, Wittich was born in Wratislavia— today Wrocław, sometimes Breslau. Changing the attribution from Brahe to Wittich unexpectedly cleared up one puzzle, explaining in a stroke why Copenhagen and Uraniborg weren't in the geographic table in Ottoboniana 1902, and why Wratislavia was.

A rich mine of Wittichian information existed in the *Tycho Brahe Opera omnia*. Its extensive index enabled us to find that Tycho had frequently mentioned Wittich in his letters. We also discovered, quite critically, that while Wittich had visited Uraniborg, a comet had appeared, and Wittich had recorded his observations in Tycho's logbook. Some years later, after Wittich had died, a countryman visited Tycho (who by then had moved to Prague), and he wrote into the logbook that he recognized Wittich's writing on the page of comet observations. We promptly wrote to Copenhagen for color photographs of those pages of the logbook, and the Royal Li-

The critical "Master Witt" annotation copied by John Craig into the margin of his De revolutionibus, *folio 82 verso, now in the University of Edinburgh Library.*

brary responded with admirable efficiency. If any doubt remained, it was totally dispelled by this new evidence. The handwritings matched beyond a shadow of doubt.

But who was the mysterious Paul Wittich? Because he had never published anything, his reputation had nearly perished. Yet, as we gradually discovered, astronomical correspondence in the sixteenth century was full of him. He was obviously very clever, mathematically gifted, and rich enough to afford at least four copies of Copernicus' book. He never settled down for long in any one place, and though he handed over some of his observations to be published by the royal physician to Emperor Rudolf II, he somehow couldn't bring himself to send any of his mathematical inventions to the printers.

One of his mathematical schemes was particularly ingenious: Wittich found how to replace multiplication and division with addition and subtraction. At first glance this sounds a lot like the invention of logarithms, a discovery attributed to John Napier of Edinburgh around 1614. In fact, Anthony à Wood, one of the great English gossips of the seventeenth century, wrote that Napier had got the idea for logarithms from a method brought back from the Continent by John Craig. In one of his copies of *De revolutionibus*, Wittich used some empty space at the end of a chapter to pen in a prototype example. He had discovered how to use the rules for sines and cosines of the sums and differences of angles to reduce multiplication of angles to addition. The prototype was rather like using a sledgehammer to crack a walnut, but at least it showed the procedure. And this was precisely one of the pages that John Craig had transcribed into his copy of *De revolutionibus* when he was being tutored by Wittich in Frankfurt an der Oder in 1576. In turn he took his annotated copy with him when he returned to Edinburgh, and he surely must have shown it to Napier, who was living in a castle in the area. Here was the paper trail to fill in Anthony à Wood's tale.

Early in the twentieth century another astronomer-sleuth, J. L. E. Dreyer, who was then producing the monumental *Opera omnia* of Tycho Brahe, had become suspicious that his protagonist had appropriated from his visitors these mathematical techniques, which go under the jaw-breaking name of "the prosthaphaeresis method."* He knew that Tycho had boasted of a handy procedure that made it easier to convert his altitude and azimuth measurements into the celestial coordinate system of latitude and longitude. Tycho had only a partial set of techniques until 1598, when he met Melchior Jöstel, a mathematics professor in Wittenberg. Dreyer, in his sleuthing through the manuscript repositories in Europe, found that, despite Tycho's claim to have worked out the method with Jöstel, the Wittenberg professor actually had the technique himself well before Tycho had arrived.

---

*Coming from the Greek, *prosthaphaeresis* literally means "addito-subtractive."

Owen Gingerich and Robert Westman signing presentation copies of
The Wittich Connection *in the Westmans' La Jolla living room, 1988.*

Dreyer, who did not know about Wittich's mathematical examples in his *De revolutionibus*, nonetheless asked, "Is not the conclusion irresistible, that similarly the invention of the method in 1580 was due to Wittich alone?" Strong evidence exists in Tycho's correspondence that Wittich had carried the *De revolutionibus* containing his mathematical methods along to Hven on his 1580 visit. A decade later, when Tycho learned that Wittich had returned to his hometown of Wrocław and had died there, he began a campaign to buy Wittich's library. In one letter he specifically asked for the three copies of *De revolutionibus* (apparently the number Wittich had brought along with him) containing the cosmological diagrams and the mathematical notes. And eventually he succeeded in buying them. So when a Jesuit librarian wrote in the Prague copy, "known to be Tycho's," he was indeed correct; the notes themselves, however, belonged to Wittich, the previous owner.

As Westman and I pored over Tycho's correspondence (published in the Dreyer *Opera omnia*), the details of the Wittich story gradually emerged. Further information on Wittich's peregrinations turned up in the correspondence of Andrew Dudith, a sixteenth-century Hungarian churchman, statesman, and devoted amateur astronomer; his original letters had disappeared during World War II, but a Czech scholar had photographed them before the war, and we managed to get a set of prints from an astronomical historian in Prague.

We made no secret of our new identification of the Prague/Vatican/ Liège annotations. Since we were bound by our agreement to publish the fruits of our researches jointly, nothing yet appeared in print, and the project was clearly too intricate for us to complete while we were still together in England. Nevertheless, I lectured rather widely on "The Mystery of Master Witt," which included the curious prehistory of logarithms. Therefore it was quite a surprise, shortly after my return to the United States, to be asked by *Sky and Telescope* magazine for an illustration from the Vatican copy of *De revolutionibus* to be used by Edward Rosen in an article entitled "Render Not unto Tycho That Which Is Not Brahe's."

For many years Rosen had had the field of Copernican studies virtually to himself, but with the Quinquecentennial he faced increasing competition. While he could be a delightful social companion, he was at heart a very scrappy New Yorker. He was also highly secretive, being plagued with fear that his researches would be plagiarized; as a consequence, he seldom shared what he was currently investigating, nor did he communicate with his colleagues to find out what they were doing. So he was in the dark about our new attribution of the pseudo-Tychonic materials. He had independently noticed that the handwriting in the Vatican Copernicus could not possibly have been Tycho's, but he had no idea who the actual annotator was.

I immediately wrote to Ed, explaining that we had known for some time that Tycho hadn't annotated the Vatican Copernicus, and that in fact, we knew the annotator was Paul Wittich, though I didn't mention our specific evidence. Rosen was outraged, as we found out afterward, because, not knowing the correct identity of the Vatican annotator, he had put into print a mistaken attribution. A stickler for accuracy, he had erred in a dozen notes in the edition of his translation of *De revolutionibus*, which had meanwhile appeared in the Polish *Complete Works* series—endnotes that attributed various readings to Brahe instead of Wittich.

Aware that Rosen was on the same trail, Westman and I rushed a note to the *Journal for the History of Astronomy* outlining the evidence for the change in attribution. But Rosen marched on as if he had been the first to assemble the evidence about Wittich and, ignoring our note, published a long article in which, among other things, he analyzed in some detail how previous biographers of Paul Wittich had gotten his death date wrong. Since we had already been onto the relevant manuscript material, we saw at once, to our amazement, that he had misread the writing. He had corrected the year appropriately but had mistaken the day, reading an abbreviation of the "5th of the Ides of January" for the "5th day of January." It was of course totally inconsequential whether Wittich had died on 5 January 1586 or on 9 January 1586, but we knew that Rosen took such minutiae very seriously indeed. We ultimately scored a trivial point in our minor game of one-upsmanship, and Ed Rosen silently corrected his error in one of his later publications.

Of more substantial importance than finding the day on which Wittich died was deducing the year in which he was born. Tycho had always implied that it was "young Wittich" who had visited him, but that could simply have resulted from his highborn arrogance. With his snobbish manners, nearly everyone seemed in some way inferior. The key to calibrating Wittich's age fell into my hands entirely serendipitously several years later, fortunately before we published our definitive account of the Wittich connection.

The library of Duke August in Wolfenbüttel was considered one of the greatest, if not *the* greatest in seventeenth-century Europe. Today Wolfenbüttel is just a small town east of Hannover, but it is still renowned for hosting one of the premier research libraries in Germany. Isaac Newton's rival, the polymath Gottfried Wilhelm Leibniz, was for some years its librarian. Among its many impressive treasures are two first editions of *De revolutionibus*, which I had examined in 1973 and again in 1978, but it was not just Copernicus' book that I was looking at in the spring of 1986. I was systematically examining *all* of their astronomy books, thinking about the possibility of a special exhibition. Among the curious volumes I opened was a printed astrology text with a set of extra pages on which the owner had constructed birth horoscopes for his classmates at Wittenberg. I quickly realized that the book had an interesting research potential. With all those birthdays I could find out how old the students at Wittenberg were. In the late Middle Ages young teenagers went to the university, but were thirteen-year-olds typically at Wittenberg in the age of Copernicus? Miriam joined me in the library and set to work with the printed matriculation list for Wittenberg University, matching names with the manuscript horoscopes. The average age of entry turned out to be seventeen, much the same as in today's universities.

Our result, published in the journal *History of Universities*, intrigued the editors so much that they included an editorial that was even longer than our article. They considered our findings to be a breakthrough in understanding the nature of the European undergraduate population in the 1500s. And our result provided the basis for deducing when Wittich was born, since we knew he enrolled at Leipzig in the summer of 1563. If he was seventeen, then his birth year was around 1546, making him the same age as Tycho. But Tycho, precocious lad that he was, had turned up at Leipzig at age fifteen. He later recalled having been at Wittenberg the same time that Wittich was there, but that he scarcely remembered him. Tycho may not have recalled that meeting very well, but his memories of Wittich's visit to Uraniborg and its aftermath were all too clear.

Paul Wittich was a cut above most of the helpers who had arrived on the island of Hven. He was more akin to Tycho both intellectually and socially. The imperious Tycho needed appreciation, and Wittich could offer it as he admired the instruments on the balconies of the Uraniborg castle. So Tycho held nothing back as he explained the novel star-sights and scales on his quadrants, sextants, and armillary spheres. They toured the library with its thousands of books and its giant celestial globe, and they swapped notes on their ingenious trigonometrical methods. And the guest showed his host the technical underpinnings of his cosmological speculations, elegantly drafted into the copies of *De revolutionibus* that accompanied his wanderings. The idea of preserving some of the Copernican details, but with the Earth at the fixed center, must have greatly intrigued Tycho. He may have already been thinking along those lines, but seeing the specific arrangement of the planetary mechanisms within a geocentric framework must surely have spurred his own imagination. Clearly, Wittich's copies of *De revolutionibus* impressed Tycho, for he spent a decade pursuing those books, specifically mentioning their number and their contents, before he finally got them.

Despite his brilliance, Wittich seems to have been much more laid-back than Tycho. I have known a number of scientists, more than competent, wonderfully helpful and full of ideas, who could scarcely ever bring themselves to turn their researches into a publishable paper. Wittich must have been one of this type, for there is a not a single published book or paper with his name on it. No doubt on Hven he became increasingly wary of the arrogant, high-intensity Tycho, and when an inheritance came his way in Wrocław, he used it as an excuse to escape the snare. He obviously accepted the huge Apianus volume, and he promised to come back, but he never did.

The next Tycho heard of Wittich was that the Wratislavian was holding forth at the observatory of his chief astronomical rival, Landgrave Wilhelm of Hesse in Kassel, revealing the secrets of Tycho's star-sights and scale graduations and much else. Tycho was infuriated and felt badly burned. Never again would he be so open about his inventions. He became a changed man, secretive to the point of paranoia. And he soon had occasion to be truly paranoid. This time it was not the upper-class Wittich, but a lowborn autodidact, once a swineherd, who became the thorn in his side.

Chapter 8
BIGGER BOOKS
LINGER LONGER

TYCHO WAS suspicious of Nicolaus Raimerus Ursus just as soon as he turned up at Uraniborg in September 1584 in the entourage of the nobleman Eric Lange. He suspected Ursus, known colloquially as "the Bear," of snooping around in his library, sniffing private papers. Determined to quench this industrial espionage, Tycho organized affairs so that the Bear got thoroughly drunk, and while he was dozing in an alcoholic stupor, Tycho had him searched. While this turned up no evidence, the lord of Uraniborg castle was still convinced that his closely guarded results were being stolen by an unwelcome guest.

And he had good reasons to worry about what Ursus had been up to, for just as Tycho was printing his own geoheliocentric cosmological system, in 1588, Ursus illustrated a very similar planetary model in a publication of his own. In both systems the Earth was at the center of the cosmos, and in both the Moon and Sun revolved about the fixed Earth, with the Sun carrying the retinue of other planets in orbit about it. However, there was a critical difference. In the Tychonic system, the circle of Mars sliced through the circle of the Sun's annual movement. This is required for Mars to come closer to the Earth than the Sun, as it correctly does in the Copernican system. In Ursus' system this was not the case, and because it always kept Mars farther than the Sun, it failed geometrically. While this error vitiated the Bear's system, it was so similar to Tycho's own proposal that the Dane was thoroughly outraged.

There was another curious feature of Ursus' *Fundamentum astronomicum* that must have galled Tycho: Ursus dedicated the many geometric

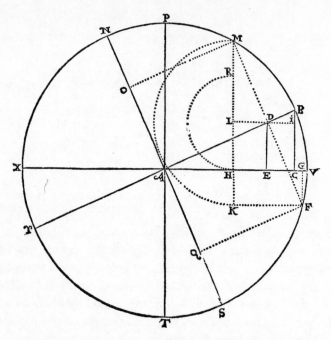

The diagram dedicated to Paul Wittich in Nicolaus Raimerus Ursus'
Fundamentum astronomicum *(Strasbourg, 1588), folio 16 verso,*
author's collection.

diagrams in his book to a veritable who's who of European astronomers. Paul Wittich got a generous two-thirds-page diagram, as did Caspar Peucer and Kepler's teacher, Michael Maestlin. Christopher Clavius, the leading Italian astronomer, had an even larger one. The biggest of all was the large folding diagram of Ursus' system, dedicated to Tycho's chief rival, Landgrave Wilhelm of Hesse. Conspicuous by his absence was Tycho Brahe. Tycho loftily declared that he was grateful not to be dragged into such a miserable book, but it still must have aggravated him to be so obviously snubbed.

Tycho wreaked his revenge on both Wittich and Ursus when he finally got around to publishing the first volume of his correspondence, where he vigorously defended his own cosmological priority. He attacked Wittich with innuendo, scarcely mentioning the name of his then-deceased one-time visitor. But Ursus, who had meanwhile become imperial mathematician at the court of Rudolf II in Prague, was depicted as a blackguard and plagiarist.

Ursus (who may well have been innocent of plagiarism) was not about to take this character assassination lying down, and he promptly prepared a scurrilous counterattack on the Danish nobleman, drawing attention to Tycho's morganatic marriage* and insinuating that his wife was a whore. As for Tycho's nose, which had been disfigured in a youthful drunken duel, Ursus slyly reported that, although he would never say so himself, one of his merry and witty dinner guests had uproariously reported that Tycho didn't actually need any instruments. He could just tilt his head up and look through his exposed nostril. "Why did he go to the senseless expense of one tool after another? He should be content with his natural nose that Mother Nature had so generously provided him with." And as for the plagiarism charge, Ursus conceded, "Let it be a theft, but a philosophical one." It was a standard rhetorical device: Concede your opponent's charge in order to demonstrate that it wouldn't have made any difference. Ursus claimed that the Tychonic system was so obvious that it was already implied in the ancient Greek work of Apollonius.

Tycho's fury was thus compounded. He brought a lawsuit against the Bear, with the consequence that the great majority of copies of Ursus' tract, *De astronomicis hypothesibus*, was destroyed. The work is now so rare that when J. L. E. Dreyer was preparing his classic biography of Tycho, he hadn't been able to see a copy (and thus didn't know that it was the bridge of Tycho's nose that had been sliced off in the duel).

Oxford English Dictionary: "The literal meaning of 'morganatic marriage' is, as explained in a 16th c. passage . . . a marriage in which the wife and the children that may be born are entitled to no share in the husband's possessions beyond the 'morning-gift'. The distinctive epithet of that kind of marriage by which a man of exalted rank takes to wife a woman of lower station, with the provision that she remains in her former rank, and that the issue of the marriage have no claim to succeed to the possession or dignities of their father."

As Westman and I were digging out the story of Tycho, Wittich, and Ursus, I had an incredible piece of luck: In the spring of 1986 I accidentally purchased a copy of *De astronomicis hypothesibus*. For many years I had been systematically collecting old ephemerides, the volumes that tabulated planetary positions on a daily basis. By analyzing the accuracy of those predictions, I could track the general lack of improvement made when the Copernican system replaced the old Ptolemaic system, for example. That was unsurprising, considering that Copernicus' achievement was not something forced by fresh observations, but rather was a triumph of the mind in envisioning what was essentially a more beautiful arrangement of the planets.

For a lover of numbers, these old volumes with their columns of digits have a compelling beauty, but this is essentially an eccentric's view—fortunately for me, because there wasn't much competition in acquiring them. And dealers knew I was the most likely buyer. Thus, when a copy of Michael Maestlin's 1580 *Ephemerides* came up for sale at the Bloomsbury Auctions in London, Quaritch's alerted me and asked if I wanted to bid. However, they warned me there was something risky about bidding because the book was completely unbound and the lot was to be sold not subject to return. In other words, caveat emptor—there was something wrong or irregular about the offering. What I know now is that if a sixteenth-century work was never bound, there is usually a reason, namely, that some part is missing.

The estimated bid was very low, but I asked Quaritch's to go higher on my behalf if necessary, because, as I explained, Maestlin's *Ephemerides* is so rare that I had never seen a copy pass through the market. Shortly after the auction Rick Watson called from Quaritch's to say that there was good news and bad news. I had won the ephemerides for less than my bidding limit, but it was a rat-nibbled and incomplete stack of leaves with a problem. "There is something else mixed in, about astronomical hypotheses," he informed me.

"That's very interesting," I responded. "Maestlin held forth in a Tübingen University disputation on astronomical hypotheses, but that little book is really rare. It would be fantastic to get it."

I suppose such occasional public disputations provided the chief intellectual entertainment for university students in those centuries. The printed versions, often not particularly attractive typographically, border on ephemera, and very few copies generally survive. Rick explained that it didn't look like a disputation, and that in any event there seemed to be a letter from Kepler dated 1595 in it. He couldn't see my jaw drop. I told him that when Kepler was a not-yet-famous high school teacher in Graz, he had written a fan letter to Nicolaus Raimerus Ursus regarding his book on the fundamentals of astronomy, and that Ursus subsequently printed it in his *De astronomicis hypothesibus.* "The book was such a fierce attack that Tycho brought legal action to have the copies banned and burned," I added, "so it's terribly rare."

"That must be it," Rick said. "The title page is missing, but Ursus' name is right here at the front."

The staff at Quaritch's hadn't checked out the lot in advance and so were completely blindsided by my extraordinary good luck. Clearly, I had ruined the week for them. But Kepler's naive enthusiasm, which had generated the admiring letter to Ursus, had pretty well ruined a year for him. He had desperately needed a new position after the Catholic rulers of southern Austria expelled all the Lutheran teachers from Graz, leaving as his best (and perhaps only) possibility to go to work for Tycho Brahe, who by 1599 had moved from Uraniborg to Prague under the patronage of Rudolf II. In fact, Kepler didn't know that Ursus had published his letter—he hadn't even kept a copy of it, and Tycho took the opportunity to give Kepler his comeuppance. The arrogant nobleman was not about to stoop to replying to a former swineherd, so instead he assigned the task to young Kepler.

Kepler took up the challenge in a typically Keplerian way. He didn't simply parry and thrust each of Ursus' jabs. Instead, he probed much deeper, into the meaning of astronomical hypotheses and the basis for deciding between them. The result was a serious philosophical treatise, unlike the polemical tracts that characterized many of the Renaissance controversies; in fact, as the translator and commentator Nicholas Jardine put it, this was the birth of the history and philosophy of science.

Kepler framed his account according to the classical rhetorical rules for a judicial oration, as he had learned at the university. Within these rigid constraints he argued that the astronomer must seek hypotheses that not only predict the phenomena accurately but are also physically plausible. These principles served well to guide his own brilliant researches.

Significant as Kepler's response was, he never saw it published. Ursus had died in 1600, shortly after Tycho had arrived in Prague and started his lawsuit. And by the end of 1601 Tycho, too, was dead. So Kepler simply stashed away the four chapters comprising his tract, where they remained among his legacy until finally, in 1858, they were printed in a comprehensive multivolume edition of his works. But there they continued to sleep for another century until Jardine took a careful look at them, and with a quotient of sideline cheerleading from me, eventually produced a scholarly analysis and English translation. While he was at it, he wrestled with the Bear's somewhat coarse Latin and included the relevant passages in his commentary.

Because of Nick Jardine's involvement with Ursus' text, I couldn't resist phoning him to share the excitement of my astonishing trophy from the Bloomsbury Auction. "My God!" he exclaimed. "You've just acquired the third known copy!"

"No, it's rare, but it can't be that rare," I protested. "There have got to be more copies than that."

It was more than ten years before another copy appeared on the market. This time Rick Watson was determined to capture it, and when he did, he researched the number of extant copies. He was able to locate only eight other copies, two in the United States (including mine) and six in Europe. I've been able to locate only three more. Tycho's lawsuit had been amazingly effective.

NORMALLY, BOOKS don't disappear so dramatically. Galileo's *Dialogo*, the book that got him in trouble with the Inquisition, was published in an edition of a thousand, and despite the ban by the Inquisitors, it remains one of the most common of the great scientific classics. Apparently, its listing in the *Index of Prohibited Books* simply made it more apt to be preserved in the seventeenth century. By the same token, Kepler was worried about sales in Catholic countries when his *Epitome of Copernican Astronomy* was placed on the *Index*, but a correspondent from Venice assured him that his book would be all the more sought after.

One of the most spectacular attritions concerns a famous English translation of Copernicus' cosmological chapters—famous because its version of the heliocentric blueprint has been so widely reproduced. In 1576 the English astronomer Thomas Digges took over his father's perpetual almanac, *A Prognostication Euerlasting*, which had already come out in six previous editions, and added to it an English version of chapters 9 to 11 of the first book of *De revolutionibus*. A large folding diagram of the heliocentric system showed the starry frame not in a spherical shell but scattered out in all directions. The caption stated, "This orbe of stars fixed infinitely up extendeth hitself in altitude spherically and therefore immovable," which was in itself a remarkable argument for the fixity of the stellar matrix. Of course this was bad news for the empyrean, the home of the blessed, for traditionally heaven had been placed right outside the shell of fixed stars. Digges's solution was ingenious. To the caption he added, "The palace of foelicitie garnished with perpetuall shininge glorious lights innumerable . . . the very court of celestial angells, devoid of greife and replenished with perfite endless joye, the habitacle for the elect."

Digges expressly stated that he had included the Copernican excerpt in the almanac "so that Englishmen might not be deprived of so noble a theory." Eventually, I stumbled onto his own copy of *De revolutionibus*, which turned up, rather unexpectedly, in the Geneva University library in the course of my systematic survey of Swiss libraries. Digges had scarcely annotated it, but he penned a telling remark on the title page: "Vulgi opinio Error" (the common opinion errs) (plate 7b). His comments thus enroll him among a handful of sixteenth-century readers who accepted the heliocentric doctrine.*

---

*It would, of course, be intriguing to find out how Digges's copy had got from England to Switzerland, but the trail has gone cold. We know only that the book arrived in what was then the Geneva Public Library in 1893 from the heirs of a local collector.

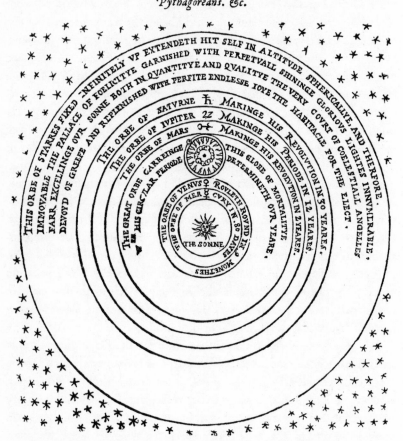

Thomas Digges's diagram of the heliocentric system, with the stars spread out toward infinity, from his A Prognostication Euerlasting *(London, 1592), author's collection.*

Thomas Digges's version of *A Prognostication Euerlasting* proved very popular and came out in eight known editions between 1576 and 1626. Typically, only two or three examples from each edition survive, so there could easily have been other editions with no known copies. Just looking over the list, I would guess that there were editions around 1581 and 1588 that have no survivors. For such popular works the production of a thousand copies seems typical, so we can estimate that about 10,000 copies were printed by 1626, yet fewer than 40 examples exist today. This is a survival rate of less than half a percent.

What happened to the 99.5 percent of those copies? Some pages were no doubt used to polish boots or candlesticks, but if unvarnished truth be told, the majority probably literally ended up as toilet paper.

Copernicus' book, on the other hand, was comparatively expensive and became famous early on, so it is unlikely that many copies were deliberately destroyed. In fact, I know of only one specific case. The large Gian Vincenzo Pinelli library, which contained a copy of *De revolutionibus*, was around 1604 being shipped by sea from Venice to Naples when it was attacked by Turkish pirates. The robbers, so disgusted at finding only books, dumped thirty-three chests overboard. Twenty-two of them were recovered and later purchased by Cardinal Federico Borromeo for his Ambrosiana Library in Milan. Because today the Ambrosiana doesn't have a copy of Copernicus' book with a Pinelli provenance, we can only conclude that Pinelli's copy went to a watery grave.

WHEN I FINALLY published my *An Annotated Census* of Copernicus' *De revolutionibus* in 2002, it described 276 copies of the first edition and 325 of the second. I am often asked how many copies were printed in the first place. Since no publisher's records remain from Nuremberg or Basel, the best I can do is make an informed guess, starting with the *maximum* number the Nuremberg printer Johannes Petreius could have printed.

Some years ago I learned that as a rule of thumb, a single sixteenth-century press could print both sides of a ream of paper in a day, that is, 480 sheets. Each sheet in *De revolutionibus* contained four pages, and there are just over 400 pages in the book. That would mean at high speed a single press could have printed 480 copies of the book in a hundred days or just under four months. The printing started in the spring of 1542, but was far from complete in the autumn of that year, perhaps because there was some delay in getting the 142 woodblock diagrams cut. The printing was finished around mid-April 1543. Thus more than seven or eight months elapsed for the main part of the printing. Printing a ream of paper a day could have produced as many as a thousand copies during those months. But is the "ream-a-day" rule reliable?

By and by I became curious about the "ream-a-day" report. Could the press really print both sides of the sheet? Wouldn't the wet ink on the one side create a mess? I consulted with veterans of hand-printing techniques, and I examined the vivid description of early printing given in Philip Gaskell's *A New Introduction to Bibliography*. Gaskell had been librarian at Trinity College, Cambridge, in the early days of my Copernicus chase, and I shared various news with him. When I demonstrated that the earliest copy of *De revolutionibus* owned by Trinity, and probably the very copy that Isaac Newton might have read when he was a fellow there, had in the meantime been sold as an imperfect duplicate, Gaskell went after it. He traded another, perfect copy with the University of Leeds, where I had found the oldest Trinity copy. It reminded me of the merchant in the *Thousand and One Nights* who offered new lamps for old!

Between Gaskell's book and the advice of the veteran printers, I discovered that the "ream-a-day" estimate was much too low. I learned that the ink had two principal components, the pigment (lampblack or soot) and the vehicle or varnish (such as linseed oil). The paper needed to be damp when printed to get the best impression, or "bite," of the type. In preparation the sheets were individually dipped in water the previous evening and stacked in a heap, generally 250 sheets. One side would be printed in the morning, and the other side in the afternoon before the sheets could dry out. Otherwise, the sheets had to be dampened again before printing on the verso, and the shrinkage could cause problems with dimensional stability. Various precautions, such as wiping off the parchment backing that touched the partially dry ink, prevented offsetting.*

Two pressmen operated the heavy press, one to manage the handheld ink balls with which he would ink up the type before the sheet of paper was put in place, and the other to slide the carriage with the type and overlying sheet of paper into the press proper and to turn down the screw to make the impression. In a really efficient operation an apprentice would place the fresh paper on the tympan (part of the folding apparatus that positioned the paper over the type), and another would remove the printed sheet and carefully stack it in the growing pile. Later the sheets would be hung up to dry. A well-run printing operation obviously had to have a capacious drying room. With a four-man operation 250 sheets could be printed on a press in an hour, About one every fifteen seconds, so 1,500 double-sided sheets could be printed in a twelve-hour day if the stamina of the pressmen held up.

Petreius no doubt had several presses, since during 1543 he produced more than twenty other titles (including three sermons by Andreas Osiander, his Copernican proofreader), and probably his typesetters alternated between projects to keep the presses at work even

---

*I assumed that everyone knew what offsetting is, but when both Miriam and my editor queried this term, I realized that my many hours spent in typesetting and my youthful passion as a collector of U.S. stamps had given me a specialist vocabulary. American stamps were originally printed from flat plates, a technique descended from the presses that Petreius used. But in 1915 rotary presses were introduced for the Bureau of Engraving and Printing, which produced the stamps. Because the process of bending the plates into a cylinder stretched the printing surface, such stamps were about a millimeter taller or wider than the flat-plate images. Rotary presses allow continuous rolls of paper to be fed into the press, a technique that made modern newspaper production possible, not to mention the long strips for coil stamps, which the post office desired. From 1918 to 1920 the postal authorities experimented with offset printing. In this process a direct image plate (as opposed to the mirror image plates of the flat-plate and rotary press plates) transferred an inked mirror image onto an "offset" roller, which then printed the actual sheet. These offset printed stamps are distinctly softer than the crisp lines of the direct printing. The mimeograph process, ditto machines, and multilith machines, commonplace in offices in the 1950s and 1960s, used this offset process so that the master sheets did not have to be prepared as mirror images. So to refer to *offsetting* means that the ink is transferred to another surface and then in turn printed onto the final sheet—in the case of a sixteenth-century press, from the back side of a partially dry page onto the parchment platen or backing sheet and then, unintended, onto the back side of the next sheet fed into the press.

Printers working with a sixteenth-century press. The pressman is inking the type while his assistant places a sheet onto the tympan and prepares to fold the frisket over it.

while a particular project had a pause for proofreading. If Petreius had used two presses, he could have printed a still larger press run in the same elapsed time, but we know he didn't do that. Petreius printed two pages side by side on each sheet, and then two more on the opposite side. Two folded sheets, one inside the other, were used to make a signature of eight pages. Petreius had a single alphabet of fancy initials used for the first letter of each chapter in *De revolutionibus*, a set cut by the Nuremberg artist Hans Sebald Beham. Only once did he need the same large letter twice on the same sheet and there the second came from a different set of letters. But if he needed the same letter in two successive signatures, as he did for the *A* in the four signatures *P* through *S*, the same fancy letter was used each time. In other words, his compositors distributed the type from each signature before they set the next one.* Had two signatures been running simultaneously

---

Distributed is another technical typographer's term, meaning that the type has been alphabetically sorted into a special type case, ready to be set again.

Johannes Petreius, printer of the first edition of De revolutionibus, *from J. G. Doppelmayr's* Historische Nachricht von den nürnbergischen Mathematicis und Künstlern *(Nuremberg, 1730).*

on two presses, he would not have had the duplicate fancy letter for the second signature. Because of the proofreading, there would necessarily have been a little lag between breaking down one frame of type and getting the next sheet ready to print. Thus it seems quite unlikely that he could have printed day after day at top speed. In seven or eight months of concentrated effort he could have printed as many as 3,000 copies, but in fact the edition was substantially smaller than this.

Although we have no records from Petreius' shop, excellent records come down to us from the sixteenth-century Plantin-Moretus Press in Antwerp. Although it was a larger establishment than Petreius' (or at least it produced more titles), it must have been comparable in many other aspects. At the well-documented Antwerp shop, the press runs ranged from two hundred (for subsidized, special editions) to about 2,500. For popular works such as liturgical texts or herbals, 1,250 was Plantin's favorite number. Since paper was one of the most expensive parts of a printing operation, Petreius would not have wanted to overestimate his sales. Clearly, a work as large and technical as *De revolutionibus* would have required a considerably smaller print run than 1,250. But how much smaller?

There are a couple of ways to use the number of surviving copies to estimate how many he actually printed, first, by comparing the number of surviving copies of *De revolutionibus* with some similar book where it is known from the records how many were printed. For example, a huge number—a thousand copies—were printed of Galileo's *Dialogo*, and a substantial number survive, so that the price of his first edition is much, much lower than for Copernicus' book. The big computer database, the OCLC,* gave eighteen copies of Copernicus' first edition, but thirty-three of the *Dialogo*. This suggests that about half as many copies of *De revolutionibus* were printed, that is, a press run of around five hundred.

Matching up numbers of books in the OCLC sounds easy, but in fact it's not, because that database is so corrupted on this kind of thing. Untrained students all over the country were hired by various colleges to enter the thousands of volumes into the database, and many smaller schools simply have facsimiles of these rare books, not the expensive treasures themselves. As a result, about a third of the purported sixteenth-century copies of Copernicus' book turn out to be twentieth-century reprints. I proposed to compare four early science titles whose press runs were known with four that were not, including the first- and second-edition *De revolutionibus*. In the end my staff had to undertake a marathon phone campaign to several score libraries in order to purge the list of the false entries.

My OCLC survey came up with three surprises. Kepler had collected part of his back salary from the emperor in paper, enough for a thousand copies of his *Rudolphine Tables*. But the scarcity of that title today suggests that only around 550 copies were actually printed (in which case Kepler sold the rest of his paper to make up for his back salary), or else that a substantial part of the press run remained unsold and was eventually pulped.

The second surprise concerned the print run of the first edition of

---

*OCLC once stood for Ohio College Library Center but was subsequently generalized to On-line Computer Library Center.

Newton's *Principia*. We know that 750 copies were printed of the second edition, but the best guess for the first edition had been 400 copies. Perhaps I shouldn't have been so astonished to find that there seem to be more copies of the first edition than the second, because the *Principia* has always sold for only about a third as much as a *De revolutionibus*. It's clearly a more common book. There is no escaping the fact that more than 600 copies of the original, 1687 edition of Newton's *Principia* were printed, possibly as many as 750 copies. By the same reasoning, substantially fewer copies of *De revolutionibus* came off the press.

Finally, the biggest anomaly of the survey was the extraordinary rarity of Galileo's first astronomical treatise, *Sidereus nuncius* or *Sidereal Messenger*, which announced his spectacular telescopic discoveries. In a letter of March 1610 to Cosimo de' Medici's personal secretary, Galileo mentioned that 550 copies had been printed. The OCLC survey picked up only five copies, compared with twenty-one for the second edition of Copernicus' book, which probably had about the same number printed. I can only ascribe this paucity to what I call Stoddard's Law: "Bigger books linger longer." (It takes half a dozen *Sidereus nuncius* copies to fill the space of one *Dialogo* or *De revolutionibus*.) I remember trying to persuade Roger Stoddard, one of Harvard's most knowledgeable librarians, to bid for a copy of *De nive sexangula* (The Six-Cornered Snowflake), a booklet that Kepler had put out as a New Year's gift for a distinguished friend, and which is now regarded as a seminal treatise on mineralogy. "It can't be too expensive," I opined, "since it's so thin." Roger groaned and retorted, "Everything you say about it makes it more expensive. Other things being equal, thin books are much harder to find than thick ones."*

ANOTHER INGENIOUS scheme for deducing the number of Copernicus copies printed, similar to the methods used by pollsters, was suggested by a Cambridge neighbor, the MIT physicist Philip Morrison. He proposed

---

*In any event, Stoddard put in a successful bid, and today Harvard has one of the world's largest holdings of Kepler titles, second only to the Württembergische Landesbibliothek in Stuttgart.

making a list of astronomers working between, say, 1543 and 1610 who would likely have owned the book, and then see how many have been found. Such a representative sample would go something like this:

Rheticus	*Mercator	*Praetorius	Muñoz
*Reinhold	Cardano	Hagecius	Tycho
Homelius	*Stadius	*Clavius	Longomontanus
Schöner	*Wittich	*Magini	Origanus
Stoeffler	*Offusius	Wilhelm of Hesse	*Maestlin
Apianus	Dee	Rothmann	*Kepler
*Peucer	*Digges	Ursus	Harriot
*Gemma Frisius	*Savile	*Schreckenfuchs	*Galileo etc.

If in my searching I had matched half as owners of the book (the asterisks in the list show the owners actually identified in the *Census*) and we assume the others on the list owned the book, but their copies have been lost, then we would deduce a survival rate of 50 percent. It's easy to see the loopholes in this procedure. Maybe not all of them actually owned the book, or maybe owners didn't bother to inscribe their books. Kepler, for example, seemed never to have written his name in the books he owned, and his copy of *De revolutionibus* was identified from other evidence. So this procedure will overestimate the number printed unless some correction factor is taken into account. If one assumes the corrected survival rate to be closer to 60 percent, then the 276 copies of the first edition and 325 of the second recorded in the *Census* would represent print runs of 400–500 and 500–550 respectively, in reasonable concordance with the larger number of printed copies of the *Principia*.*

If my estimate of the number printed is correct, then more than half the copies printed have survived. Unlike the ephemeral *A Prognostication Euerlasting*, Copernicus' *De revolutionibus* quickly gained a reputation as an important book, so few people would have deliberately destroyed a copy,

---

*It might be objected that if the 1543 *De revolutionibus* sells for two or three times as much as the 1687 *Principia*, it ought to be two or three times as rare. The numbers do not scale this way because Copernicus' book is just enough scarcer that the pressure on its price is actually quite disproportionate to its rarity.

although it's appalling to remember that the entire Oxford University Library was sold for scrap in the mid-1500s. Nor was that situation unique to Oxford, as libraries were deconstructed throughout the land. The radical English Protestant reformer and sometime playwright John Bale, writing in 1549, remarked that the purchasers of libraries "reserved of those lybrarye bokes, some to serve theyr jakes [johns], some to scoure theyr candelstyckes, & some to rubbe their bootes. Some they solde to the grossers and sopesellers, & some they sent oversee to the bokebynders, not in small nombre, but at tymes whole shyppes full. . . . I know a merchaunt man, whych shall at thys tyme be namelesse, that boughte the contentes of two noble lybraryes for xl shyllynges pryce, a shame it is to be spoken."

The fact that *De revolutionibus* was a fairly expensive book may have helped protect it. Here and there I found copies with prices written in, although the currency is apt to be ambiguous. The best record is in a copy I found in Dresden, where the astronomer Valentin Engelhart in 1545 tallied the prices of several titles bound together. The Copernicus cost one florin, equivalent to twelve groschen. During this time the university matriculation fees were in the range of six to ten groschen, and when Rheticus was being enticed to a professorship in Leipzig, he was offered the special salary of 140 florins per annum. I have wrestled with the value of currency in the sixteenth century for a long time. Personal services and food were cheap in those days, so the relative strength of the money can't easily be compared with living standards today. Perhaps an astronomy professor who took his studies seriously didn't balk at paying about 1 percent of his annual salary for an important volume, but he surely took good care of his book.*

So, WITH 400–500 copies printed in 1543, and 276 accounted for in the *Census*, where did the missing Copernicus books go? Inspecting hundreds

---

*The size of the sixteenth-century library of one Stephan Roth is astounding. He had studied under Luther in Wittenberg and eventually became chief city clerk in Zwickau. I went to see his Copernicus in the summer of 1976 in what was then a drab East German town, but the library was wonderful. Roth had amassed a collection of 6,000 titles in the 1530s and 1540s. Being city clerk must have been a lucrative post!

of copies of *De revolutionibus* has convinced me that water is the chief enemy of books. A substantial number show the traces of dampstaining. For millennia architects have been working to perfect roofs, but every time I visit Harvard's Science Center during a rainstorm, I know it has been a losing battle. And when I served on the Yale Library Visiting Committee, our first assignment was a tour to see how disastrously the roof of the Sterling Library leaked. For every heavily dampstained copy recorded in the *Census*, there must have been a copy discarded because it was so thoroughly soaked that it turned purple with mildew or simply became papier-mâché.

Bookworms have riddled a number of copies. I thought I had never laid eyes on a bookworm, living or dead. Many of my students suppose it's a mythical beast and are incredulous when shown pages perforated by their trails. I didn't remember even seeing a picture of a bookworm, so was taken by surprise to discover that one is shown in Robert Hooke's well-illustrated *Micrographia* of 1665. From the small, round bores in early books, I had always assumed that the hungry insect was a cylindrical worm, but Hooke's enlargement pretty clearly shows a silverfish. Hooke himself described the insect as "the silver-colour'd Book-worm" and reported that "this Animal probably feeds upon the Paper and covers of Books, and perforates them in several small round holes." In fact, the *Encyclopaedia Britannica* indicates that a variety of insects qualify as bookworms, with the silverfish (*Lepisma saccharina*) as the leading candidate.

I was pretty puzzled about how a silverfish could create round bore holes, but eventually found the answers from Nicholas Pickwoad, an English expert who was helping Harvard University on its numerous book conservation problems. The silverfish feasts on mold damage, so it proliferates in humid environments. Its damage is generally to the surface of a page or to a leather binding. In contrast, the round bore holes often seen in early books were caused by the hungry grubs of the deathwatch beetle (family Anobidae), which can eat right through the pages of books on a library shelf or penetrate furniture. The beetle lays it eggs near a source of food, for instance, in a crack or crevice of a well-stocked bookshelf, and the larva bores its way through its food supply, sometimes taking as long as ten years before it finally metamorphoses into a beetle.

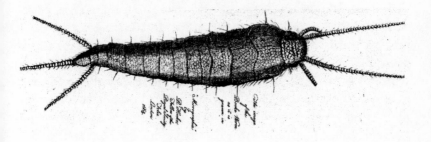

The silverfish bookworm from Robert Hooke's Micrographia *(London, 1665).*

Presumably, the really well-drilled copies of *De revolutionibus* have long since been scrapped.

Rodents can make even quicker work of an ill-fated volume. A few years ago the Carnegie Institution of Washington put its library in warehouse storage while its premises were being remodeled. The books were placed on sledges and carefully covered with tarpaulins to secure them against water damage; not until several weeks later did it finally occur to someone to include rat poison in the precautions, but it was already too late. Today the Carnegie Institution has a library with scores of missing spines nibbled away by the rodents.

Fire ranks low in the list of book destroyers. I have tried without success to document whether any copies of Copernicus' book were lost in the Great Fire of London in 1666. Possibly so, but there is no evidence. A copy was lost when the Great Tower burned in Copenhagen in 1728, presumably another when the Strasbourg Library was destroyed in the Franco-German War in 1870, and a first edition went up in flames when the retreating Nazis deliberately burned the National Library in Warsaw in the autumn of 1944. Demolition bombing in World War II brought about substantial losses of *De revolutionibus*, in Douai, Frankfurt, Munich, and Dresden.

THE MORE cheerful reverse side of the coin is not the enemies of books but their friends, the booksellers who made it possible to obtain a copy in the first place. When *De revolutionibus* was published in 1543, printing with movable type had been in use less than a century. Yet by that time modes of international distribution had already been established, primarily through large regional fairs. Printers and booksellers came together especially in Frankfurt, which had already been established as a leading fair in the late Middle Ages. In 1543 it would have been the place for booksellers to pick up a stock of Copernicus' book, always as stacks of unbound sheets.* Being a bookbinder was an entirely different profession from the printer; not until the seventeenth century did it become relatively common for printers to package their wares in cheap temporary paper bindings.

In 1564 a catalog listing the various new titles on offer at the Frankfurt Book Fair was issued, the first in a long-running series of such booklets. Toward the end of the century, when young Johannes Kepler proudly authored his first major book, it too was offered at the fair—but Kepler was devastated to discover that his name was misspelled in the catalog. I didn't have much sympathy for Kepler when I read about this, because he tended to write the *K* of his name so ambiguously that it could easily be mistaken for an *R*.† In any event, I became very curious to see a copy of the catalog in question.

One day when I was in Harvard's Houghton Library, it occurred to me to inquire how to find the catalogs in a major German library such as Munich's or Stuttgart's. I knew that the European libraries generally listed their books just by authors, so I wondered whether I would have to know the cataloger's name to find out if the old Frankfurt Book Fair catalogs were held there. I had forgotten about my inquiry when roughly a year later the senior librarian and I bumped into each other in the reading room once again.

---

*Even today the annual Frankfurt Book Fair, held in the fall, brings publishers together from all parts of the world.

†Ursus had mispelled the name as Repler when he seized the opportunity of including Kepler's letter in his scurrilous attack on Tycho Brahe.

"Do you remember your inquiry about the Frankfurt Book Fair catalogs?" he asked. I nodded affirmatively. "Well, we've found them."

I was rather taken aback by the way he put it. "What do you mean, you've *found* them?" I asked.

He hung his head, at least figuratively, and explained that the library had recently got an inquiry from someone who wanted to produce a facsimile of one of them. It turned out that just before World War II Harvard had bought a leading collection of these catalogs, but they were so complicated bibliographically that no one had had the courage to catalog them, so they just sat in the Houghton stacks until everyone completely forgot about them. The staff finally located the bundle after they realized that they had to be there someplace.

In the early 1970s a German publisher made a facsimile edition of the entire set, except that the catalog for the autumn fair in 1598 couldn't be found anywhere in Europe. Researchers finally traced Harvard's purchase, and there it was, probably the unique copy of the missing catalog. ("Little pamphlets linger least.") It is fascinating to pore over the thin, fragile catalogs, a spring and autumn edition each year, with the books arranged in broad categories, the Latin titles in Roman letters and the German titles in Gothic type: Protestant theology, Pontifical theology, Music, History, Poetry, and "Philosophy, Humane Arts, and Polite Literature." In that latter category in the fall of 1566, the third year of the catalogs, Copernicus' *De revolutionibus* was listed. It was a folio from Basel, publisher and price not specified. This, the second edition, came from Heinrich Petri, a printer of scholarly works in Basel, possibly a relative of Johannes Petreius of Nuremberg. By 1566 Heinrich had involved his son, Sebastian, the man who would eventually carry on the business under the name Sebastian Heinrichpetri, so in that year the publisher's identification on the title page read "Ex officina Henricpetrina."

The supreme irony of this story is that the Harvard set of Frankfurt catalogs isn't complete. Despite having the unique 1598 catalog, Houghton Library lacks the spring and fall 1597 catalogs, one of which listed Kepler's *Mysterium cosmographicum*. I now know you have to look under Georg

Willer to find them, but I still haven't actually seen an original copy listing the author Repleo.

INCIDENTALLY, THERE was another way books could disappear—by being deliberately destroyed when they were listed in the *Index of Prohibited Books*. *De revolutionibus* was placed on the Catholic *Index* in 1616.

Chapter 9

FORBIDDEN GAMES

GEORG JOACHIM RHETICUS needed many months to persuade Copernicus to send his book to the printer. The Polish astronomer's reluctance to publish his manuscript must have arisen from a complicated jumble of reasons and phobias. In the first place, there was no suitable printer nearby to handle a complex technical work, and besides, there were still many details not quite polished to his satisfaction. But lurking in the background was the fear that his ideas about the mobility of the Earth, so contrary to common sense, would lead to his being hissed off the stage. That is precisely how he expressed it in the dedication that he eventually wrote to Pope Paul III.* And he knew that he might be in trouble with some religious sensitivities. Though he explicitly dismissed this in his dedication to the pope, saying, "Perchance there will be babblers who take it on themselves to pronounce judgment, although completely ignorant of mathematics, and by shamelessly distorting the sense of some passage in the Holy Writ to suit their purpose, will dare to find fault and censure my work, but I shall scorn their attack as unfounded," he must have worried about their potential criticism.

---

*Copernicus used the word *explodendum*, which means "being hissed or clapped off the stage." The *Oxford English Dictionary* confirms that this is also the original but now obsolete meaning of the English word *explode*, which did not pick up the modern definition of "to blow up with a loud noise" until around 1700. Shakespeare never used the word despite its theatrical connotation, but his contemporary Kepler did, undoubtedly in an echo of Copernicus' usage, when in the introduction to his *Astronomia nova* he wrote (in Latin), "First, Ptolemy is certainly hissed off the stage." Kepler may have been sensitized to the word by Galileo, who used it in his first letter to Kepler, in 1597.

Foremost among the biblical verses that could cause trouble was a vivid passage from the Book of Joshua.

> On that day when the Lord delivered the Amorites into the hands of Israel, Joshua spoke with the Lord, and he said in the presence of Israel:
>
> > *"Stand still, O Sun, over Gibeon*
> > *and Moon, you also, over the Valley of Aijalon."*
> > *And the Sun stood still and the Moon halted,*
> > *till the people had vengeance on their enemies.*
>
> Is this not written in *The Book of the Just*? The Sun stood still in the middle of the sky and delayed its setting for almost a whole day. There was never a day like that, before or since.

In Wittenberg, even before *De revolutionibus* was printed, Martin Luther had cited the Joshua passage in the course of a dinner conversation. Apparently, Luther had heard about the new cosmology from Rheticus or Reinhold at the university. Despite the informality of the mealtime setting, an eager student named Anton Lauterbach copied down the critique: "There was mention of a certain new astrologer who wanted to prove that the Earth moves and not the sky, the Sun and the Moon. This would be as if somebody were riding on a cart or in a ship and imagined that he was standing still while the Earth and trees were moving. Luther remarked, 'So it goes now. Whoever wants to be clever . . . must do something of his own. This is what that fellow does who wishes to turn the whole of astronomy upside down. . . . I believe the Holy Scriptures, for Joshua commanded the Sun to stand still, and not the Earth."

Or maybe that's not exactly what he said, because another student, Johannes Aurifaber, later reported it a little differently. "That fool would upset the whole art of astronomy," Luther supposedly said, in what is one of his most widely quoted lines, though the experts generally believe this version is apocryphal if for no other reason than that Aurifaber wasn't actually present at the dinner.

These off-the-cuff remarks might have been forgotten, though they were printed along with numerous other of Luther's conversational views in the *Tischreden*, or "Table Talk," series, first published in Wittenberg in 1566, long after Luther's death. But his opinion was given a popular press when Andrew Dickson White, the first president of Cornell University, publicized the reformer's Copernican remarks in 1896 as part of his *A History of the Warfare of Science with Theology in Christendom*. A liberal Christian, White announced that his goal was to let "the light of historical truth into that decaying mass of outworn thought which attaches the modern world to medieval conceptions of Christianity—a most serious barrier to religion and morals." He was eager to discredit what he believed was religion's antipathy toward the march of science, so he got his graduate students to dig up as many cases as they could find. The so-called Galileo affair played a central role in his account, introduced by the following wholly fictitious episode.

> Herein was fulfilled one of the most touching of prophecies. Years before, the opponents of Copernicus had said to him, "If your doctrines were true, Venus would show phases like the moon." Copernicus answered: "You are right; I know not what to say; but God is good, and will in time find an answer to this objection." The God-given answer came when, in 1611, the rude telescope of Galileo showed the phases of Venus.

It is true that Galileo's telescope revealed for the first time that Venus went around the Sun, contrary to the Ptolemaic arrangement, but neither Copernicus nor his opponents ever considered such a test. The seeds for this myth were planted, perhaps inadvertently, by the English astronomer John Keill in a Latin textbook he published in 1718. With each retelling the story was more richly embroidered, reaching its apotheosis with White's well-embellished vignette.

In any event, since the infamous Galileo affair played out in a Catholic setting, White was eager to give Protestants a role, too, in religion's presumed resistance to the advances in scientific knowledge. This is why he dusted off a version of Luther's comment. But the former Cornell president was not about to stop with Luther. Despite the fact that Copernicus' book was essentially published under Lutheran auspices, as was the first volume of tables based on *De revolutionibus*, White continued, "While Lutheranism was thus condemning the theory of the earth's movement, other branches of the Protestant Church did not remain behind. John Calvin took the lead, in his *Commentary of Genesis*, by condemning all who asserted that the earth is not at the centre of the universe. He clinched the matter by the usual reference to the first verse of the ninety-third Psalm,* and asked, 'Who will venture to place the authority of Copernicus above that of the Holy Spirit?'"

No doubt White's quotation from Calvin increased the readership of Calvin's works, for it set historians of science off on a frustrated search to find where the Genevan reformer mentioned Copernicus. In 1960 Edward Rosen, a master of minutiae, not only tracked down a flock of authors who simply parroted White's account, but traced the comment itself back to the Reverend F. W. Farrar, an Anglican canon who was at one time chaplain to Queen Victoria and who overconfidently relied on his capacious memory of quotations to generate out of whole cloth Calvin's comment on Psalm 93. In a sweeping generalization Rosen concluded that Calvin had never heard of Copernicus and therefore had no attitude toward him.†

Given the relatively wide distribution of *De revolutionibus*, I think it highly likely that John Calvin saw the book, but he probably assumed from the notice on the back of its title page, addressed "To the Readers Concerning the Hypotheses in this Book," that Copernicus' book was intended as a mathematical device for calculation and not a real description of nature. This *Ad lectorem* declared that "it is the duty of an astron-

---

*"Thou hast fixed the Earth immovable and firm, thy throne firm from of old; for all eternity thou art God." †There the matter stood for a decade. Then, in 1971, a French scholar noticed that in a sermon on I Corinthians 10 and 11, Calvin denounced those "who will say that the Sun does not move and that it is the Earth that shifts and turns." Calvin neither mentioned Copernicus by name, nor did he invoke any Scripture against heliocentrism itself. In fact, it has been cogently argued that Calvin was alluding to a quotation in Cicero brought on by a debate with one of his understudies who had fallen out of his good favor. So the jury is still out on Calvin's opinion, if any, on Copernicus and his book.

omer to record the motion of the heavens with diligent and skillful observations, and then he has to propose their causes or, rather, hypotheses, since he cannot hope to attain the true reasons. . . . Our author has done both of these very well, for these hypotheses need not be true nor even probable; it is sufficient if the calculations agree with the observations." It was added by Andreas Osiander, the learned theologian-minister of the Sankt Lorentz Kirche in Nuremberg who proofread most of its pages. When Copernicus' friend Bishop Tiedemann Giese saw this unauthorized addition, he was greatly exercised and wrote a letter to the Nuremberg City Council demanding that the front matter be revised and reprinted. To Copernicus' first and only disciple, Georg Joachim Rheticus, he expressed the wish that in the copies not yet sold, there should be inserted the little treatise "by which you have skillfully defended the idea that the motion of the Earth is not contrary to the Holy Scriptures." Such a replacement would have been relatively easy at a time when books were sold unbound, as loose stacks of folded signatures.

Because of Bishop Giese's letter, Copernican scholars had long known that Rheticus, in addition to writing the *Narratio prima*, or "First Report," that served as a trial balloon for the radical heliocentric cosmology, had also written another tract discussing how to understand those scriptural citations that seemed at odds with a moving Earth. But none of his contemporaries, nor, for that matter, Andrew Dickson White and his students, ever knew what Rheticus, a staunch Lutheran, had written on heliocentrism and Holy Scripture. For many years it was assumed that Rheticus' report had been lost in the dustbin of history. Then, almost miraculously, it was rediscovered in the flurry of Copernican researches associated with the 1973 quinquecentennial year.

It turned out that Rheticus' reconciliation of Copernicus' science and Scripture had actually been printed, but anonymously, in a little booklet published in Utrecht in 1651. This long-overlooked tract, now apparently existing in only two copies,* was identified and described by the Dutch

*One copy is in the British Library, and the other is in the library in Greifswald, Germany.

historian of science Reijer Hooykaas. Thus we now know that Rheticus quoted Augustine as saying that Scripture borrows a style of discourse from popular usage "so that it may also fully accommodate itself to the people's understanding and not conform to the wisdom of this world." Rheticus emphasized this point repeatedly. He cited a series of passages commonly used to condemn the reality of the heliocentric plan, including Joshua and the battle of Gibeon, and noted that things appear to move either because of the motion of the object itself, or because of the movement of one's vision, but that common speech mostly follows the judgment of the senses, that is, the appearance that the motion is in the object itself. "As persons who seek the truth about things," he wrote, "we distinguish in our minds between appearance and reality."

Yet, decades later, the same scriptural passages were still being used in their literal sense. And, long before Rheticus' little tract was printed (over a century after it was written), his arguments had been independently discovered and advocated by two of the leading Copernicans, Kepler and Galileo. Both Kepler's and Galileo's copies of *De revolutionibus* survive, preserved in European libraries, and in both cases their annotations tell part of the story of the religious reception of this epoch-making book.

I FIRST SAW, and photographed, Kepler's copy in the Leipzig University Library in the summer of 1972, a year before the quinquecentennial celebrations. In those days one needed a sponsor to enter East Germany, and Miriam and I found one in the publisher Edition Leipzig. It provided the entrée into that fenced-off police state and enabled me to see rare books in several libraries. But the most memorable, and spookiest, part came as we were driving out of the country. We had easily cleared the potentially troublesome East German customs—worrisome because I was carrying out undeveloped film—but the young officer had been more interested in practicing English than in searching our luggage. As darkness fell, a large truck blocked the view of the actual exit, and we unwittingly headed off onto a stretch of abandoned Autobahn that meandered through the no-man's-land between East and West Germany. Some distance along this dim and suspiciously untrafficked route we realized that grass was growing down the middle of the highway. In stark terror we did a U-turn, hoping no border guards were taking aim at us. Nowadays it's hard to recapture the sense of relief that always came during those cold war times when one finally transited back into the free world.

And so I carried out the latent images of Kepler's *De revolutionibus*. Some of the most distinctive features in the book are not the notes he wrote in it but what was already present when he acquired it. The copy had originally been given by the Nuremberg printer Petreius to a local scholar, one Jerome Schreiber. Schreiber was clearly an insider, and he learned who had written the anonymous *Ad lectorem*, the advice to the readers printed on the back of Copernicus' title page. One of the slides I made in Leipzig shows the name Osiander written by Schreiber above the *Ad lectorem*. That annotation tipped off Kepler to the unnamed author's identity.

Kepler was particularly incensed by this anonymous introduction, because, unlike the great majority of sixteenth-century astronomers, he was a realist, and he believed that Copernicus, too, thought that the heliocentric system was a real description of the planetary system and not just a mathematical computing device. Hence he was pleased with himself when he could put his own advice to the reader on the back of the title page of his great *Astronomia nova*, which was published in 1609 and in which he presented the evidence that the orbit of Mars was not a perfect circle but an ellipse. With the notice on the back of his own title page, Kepler revealed in print for the first time that Osiander had authored the *Ad lectorem*. He made it clear that Copernicus would not have subscribed to it. Osiander's advice was that the cosmology in the book was merely hypothetical, that "perhaps a philosopher will seek after truth but an astronomer will just take what is simplest, and neither will know anything certain unless it has been divinely revealed to him." Because this advice had essentially protected the book from religious condemnation for many decades, Kepler's revelation that Copernicus had not written nor subscribed to that caveat was dynamite; it undermined the Church's position that heliocentrism was a strictly hypothetical scheme, useful for mathematicians but not to be confused with physical reality. Thus the little notice behind the title page of the *Astronomia nova* helped set the stage for the prohibition that would soon follow.

I first saw Galileo's copy, a second edition in the National Library in Florence, in July 1974. I found myself disbelieving that the book had really belonged to the Italian astronomer, for this copy had no technical marginalia, in fact, no penned evidence that Galileo had actually read any substantial part of it. Yet, as I finally convinced myself, his handwriting was there. He had carefully censored the book according to the instructions issued from Rome in 1620—universal instructions that he had helped trigger.

Galileo's censored Copernicus was just one highlight of an intense weeklong research expedition in north-central Italy. Another memorable moment came in Padua, where the local archives agreed to let me photograph the oldest dated autograph document in Copernicus' own hand.

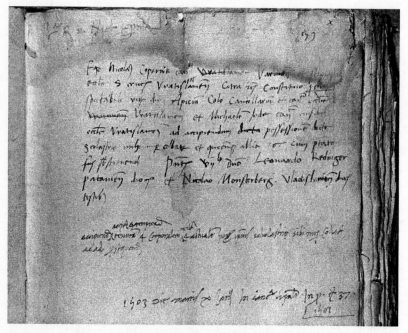

Copernicus' earliest dated signature, 10 January 1503, from the Padua State Archives.

I had brought along my case of photoflood lamps and found a precarious place near an electrical outlet where I could both clamp the lamps and place the open book of documents.* Copernicus, while a medical student in Padua in 1503, had written out a statement to be notarized, and that was the manuscript I photographed.

In twentieth-century Central Europe, bitter contention broke out concerning Copernicus' ethnic origins, which became particularly shrill during the Nazi period. The Germans argued that the astronomer's family name was a Germanic Coppernigk or Koppernig, whereas the Poles defended Copernik. On the Padua document his signature clearly reads Nicolaus Copernik, though he was rather indifferent about orthography and later would occasionally sign off as Coppernicus.†

Besides more than a dozen Copernicus volumes recorded on the Italian field trip, Miriam and I listened to sonorous Armenian chorales in the mosaic-laden cathedral in Ravenna, saw *Coppelia* danced by the La Scala ballet company in Milan, and feasted our eyes on the Botticellis, Fra Angelicos, and Michelangelos in Florence and the Giottos in Padua. And while we were in Padua I also photographed the Renaissance anatomical theater, one of only two original surviving examples, the other being in Uppsala. Here medical students in Copernicus' day—and probably Copernicus himself—crowded into the ranks of the small oval auditorium to observe human dissections. Typically, a barber-surgeon would do the cutting while the professor read the relevant text from Galen. Quite possibly, the Italian penchant for cutting up saints to serve as holy relics helped make the medical dissections socially acceptable. We got the flavor of this when we visited the library in Ferrara (which had a censored copy of *De revolutionibus*) and were there astonished to discover Ludovico Ariosto's heart in a large urn,‡ and we found Galileo's index finger in a reli-

---

*Despite the makeshift arrangement, the photograph was good enough to appear in the appropriate *Complete Works* volume of the Polish Academy of Sciences.

†Such indifference was characteristic of the period: Kepler sometimes spelled his name Keppler.

‡The Michelin Guide characterized Ariosto's *Orlando Furioso* as the Renaissance equivalent of *Gone with the Wind*.

quary in the Florence History of Science Museum (whose *De revolutionibus* was not censored).

Although Galileo made his most significant astronomical discoveries in Padua—the mountains on the Moon, the myriad stars composing the Milky Way, the satellites of Jupiter—his years of fame and confrontation with the Church took place in Florence. Like Kepler, Galileo endorsed a realist stance, and to this end he wrote a privately circulated letter in which he argued (as Rheticus had done, though he didn't know this) that the Bible spoke in idiomatic terms so that everyone could understand it, but it was not a scientific textbook. His punch line was that "the Bible teaches how to go to heaven, not how the heavens go." It was one thing for Kepler, who had been trained as a Lutheran theologian, to argue this way in a Lutheran context in the introduction to his *Astronomia nova*, but quite another for Galileo, who was at best an amateur theologian, to argue the same way in a Catholic context.

In 1545 Pope Paul III had convened the Council of Trent to discuss both Church reform and a hardened line against heretics, and following Pope Pius IV's publication of the council's decrees in 1564, the Protestant protest was no longer seen as an in-house debate. Catholic scriptural interpretation was reserved for the trained hierarchy, and Rome did not like the idea of an amateur theologian rocking the boat when they had their hands full with the heretical dissidents north of the Alps. If Galileo was invading its interpretive turf, the Vatican determined to remind everyone just who had authority by banning Copernicus' book. But it was not that easy. The Vatican claimed the power to establish the calendar, as it had done with the Gregorian reform in 1582. A fundamental part of calendar reform was specifying how to determine Easter, and this required astronomical knowledge of the Sun and Moon. *De revolutionibus* included observations of the Sun and Moon, of potential value to the Church, so it was inadvisable to ban the book outright. Nor could the heliocentrism simply be excised, for it was too firmly embedded in the text. The only path was to change a few places to make it patently obvious that the book was to be considered strictly hypothetical.

Thus it happened that in 1616 *De revolutionibus* was placed on the Roman *Index of Prohibited Books* "until corrected," and in 1620 ten specific corrections were announced. Two examples will show what the Vatican was up to. The heading of Book I, chapter 11, "The Explication of the Three-Fold Motion of the Earth," was changed to "The Hypothesis of the Three-Fold Motion of the Earth and Its Explication." At the end of the preceding chapter, where Copernicus had declared that the great extent of the starry universe made it impossible to detect any annual oscillation in the positions of the stars owing to the Earth's annual motion, he had exclaimed, "So vast, without any question, is the divine handiwork of the Almighty Creator!" This, too, met with the censors' condemnation. Why was such a pious statement forbidden by Rome? Simply because it made it appear that God had created the universe from a heliocentric perspective. All of these corrections were dutifully recorded by Galileo in his copy of the book, though he took care to cross out the original text only lightly so that he could still read it.

Because the Vatican Congregation of the Index was so explicit about its instructions, it is almost immediately obvious whether or a not a copy was censored, something I took note of as I examined and reexamined several

The Inquisition's censorship of De revolutionibus, *in Galileo's hand in his personal copy.*

hundred copies of the book. In addition, I systematically recorded provenances, that is, where the books had been. For about half of the copies it was possible to determine where they had been in 1620 when the instructions for censoring the book were announced. Then a very interesting result emerged, something the Inquisitors never knew. Roughly two-thirds of the copies in Italy were censored, but virtually none in other countries, including Catholic lands such as Spain and France. It became apparent that the rest of the world looked on the exercise as a local Italian imbroglio, and they were having none of it. In fact, the Spanish version of the *Index* explicitly permitted the book!

Copernicus' *De revolutionibus*, and two later additions, Kepler's *Epitome of Copernican Astronomy* and Galileo's *Dialogue on the Two Chief World Systems*, remained on the *Index* until this became something of a scandal. Not until 1835 did a copy of the Roman *Index* appear without these titles.*

FOR TWO YEARS during the 1980s I hosted a student from mainland China at the Harvard-Smithsonian Center for Astrophysics. Shida Weng was a victim of China's cultural revolution, consigned to the backwaters when he might profitably have been in graduate school, but he had spent the time teaching himself English. In 1982 the Chinese government sent him to Cambridge to study the history of astronomy. During his stay I took Shida as my guest to the History of Science Society meeting in Norwalk, the one where Bob Westman and I presented our results on Wittich's annotated copies of Copernicus' book. En route I asked Shida what had been his most surprising impression of the United States. He replied that in America we have so many choices, what to wear, what to eat, what to read—a perceptive analysis, I thought. I told him that I hoped to visit his country someday.

---

*Incidentally, Andrew Dickson White, though an unreliable guide to the religious reception of Copernicus' book, did own a first edition of *De revolutionibus*, now preserved in the library of Cornell University. A previous owner had been the English antiquary John Aubrey (1626–97), and since the book had spent the critical years in England, it was not censored.

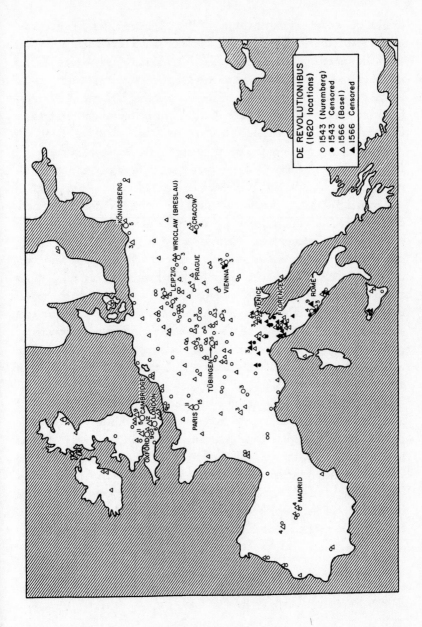

DE REVOLUTIONIBUS
(1620 locations)

○ 1543 (Nuremberg)
● 1543 Censored
△ 1566 (Basel)
▲ 1566 Censored

KÖNIGSBERG

WROCLAW (BRESLAU)
CRACOW

LEIPZIG

PRAGUE

VIENNA

VENICE
FLORENCE

ROME

PARIS

TÜBINGEN

CAMBRIDGE
LONDON

OXFORD

MADRID

Our opportunity came in the fall of 1985. Miriam and I decided to begin an Asian tour in China, and Shida provided a timely invitation. Besides the mandatory visit to the Great Wall and a trip to see Xi'an's astonishing terra-cotta army, I had two specific goals: I wanted to see the Pascal calculating machine sent to Beijing possibly by Louis XIV, and I hoped to inspect the two copies of *De revolutionibus* brought to China by Jesuit missionaries in 1618.

For centuries astronomers were the major consumers of numbers. In 1623–24 Kepler's friend Wilhelm Schickard (whose well-annotated *De revolutionibus* is today in the Basel University Library—see plate 7e) designed a calculating machine to assist in Kepler's continuing numerical problems, but unfortunately it was destroyed in a fire before it could be tested in real calculations. The oldest surviving mechanical calculating machine, from 1644, was made by the French mathematician Blaise Pascal, not for astronomers but to assist his father with financial calculations at the customs office, and a few original examples survive.* In the seventeenth century wealthy Europeans coveted the silks and porcelain that came only from China, but the Chinese were not interested in very much of what Europeans had to offer. One exception was fancy clocks, and another was automata, such as mechanical dolls that danced to the tunes of mechanical music boxes.

The Pascal calculating machine was a novelty of precisely the sort to intrigue the Chinese court, and in researching the history of automatic calculating devices for a Charles Eames exhibition, I had learned that there were two Pascal machines in the Royal Palace in the Forbidden City in Beijing. But even when Miriam and I arrived in Beijing, Shida Weng was not at all certain that we would be given permission to see the Pascal machines, which were in a restricted section of the Royal Palace. Then, as mysteriously as many things in China, the permission was

---

*Several examples are found in the Conservatoire des Arts et Métiers in Paris, another in the historical museum in Clermont-Ferrand (Pascal's hometown), one in the State Mathematical-Physical Salon in Dresden, and one in the IBM Collection in New York.

granted. We were taken into rooms not ordinarily open to visitors, and there the devices were gently lifted out of their packing boxes. One proved to be the genuine French import. The other, we were surprised to discover, was an almost exact Chinese copy. Later, when I reported our experience to Joseph Needham, the eminent English student of Chinese science and civilization, he allowed that we were probably the only westerners to have examined these machines in the past half century.

The two copies of *De revolutionibus* arrived in China about thirty years earlier than the Pascal calculator. Shida Weng himself had never before gained permission to examine these books, and seeing them was only slightly less of a cliff-hanger than viewing the Pascal machines. We were taken to the National Library, which must have been a very suitable and handsome structure when it was built in 1931, but was in 1985 (like many libraries around the world) awaiting more commodious quarters.* The rooms were filled with readers save for the one reserved for the works of Marx and Lenin, which was as quiet as an undiscovered tomb. The rare book room was chilly and singularly austere, but three volumes I had specifically requested were on reserve for us: a second edition of *De revolutionibus* and also a third edition from 1617, both of which were brought by Jesuit missionaries in 1618; and a copy of Kepler's *Rudolphine Tables*, which obviously came somewhat later since it wasn't published until 1627 (plate 8a). The Jesuit library was not nationalized until after the Chinese revolution of 1949. In this action the Chinese were singularly backward, since most of the Jesuit libraries in Europe had been nationalized a century and a half earlier.

The two editions of *De revolutionibus* had traveled to China at an especially interesting time, because the Vatican decree prohibiting Copernicus' book "until corrected" had appeared in 1616, but the corrections were not issued until 1620, by which time the two Jesuits who had brought the books had already made their way to Beijing. The 1566 edition, brought by Giacomo Rao, included Rheticus' *Narratio prima*, which had been

---

*A new building was opened in October 1987; it is now the largest single library building in the world.

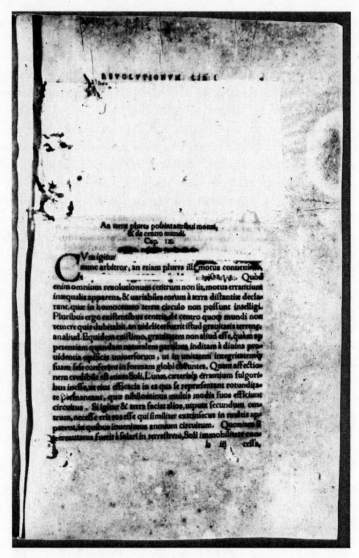

The draconian censorship of the entirety of Book I, chapter 8, the only known copy in which the more severe form of the Vatican's excisions was performed. Biblioteca Statale, Cremona.

reprinted as an appendix; Rao crossed off Rheticus' name on the title page because as a Lutheran Rheticus was considered a "first-class" author, meaning that all of his works were banned, but the *Narratio prima* was not excised (as sometimes happened). However, the Lutheran names at the beginning of the appendix were struck out. Otherwise the book was uncensored and unannotated.

The third edition, brought to China by Nicholas Trigault, S.J., was treated slightly differently. Chapter 8, on the refutation of the ancient arguments against the mobility of the Earth, was marked in Latin, "This chapter is not to be read." This must have been a shrewd guess, probably on Trigault's part, because among the remarks contained in the 1620 decree was the statement that chapter 8 could be excised in its entirety. The decree went on to say, however, that because students might wish to see how the arguments of the book were developed, it would be satisfactory simply to reword two of the passages in the chapter. Presumably, this instruction never caught up with Trigault. Among the nearly six hundred copies examined in my survey, only one suffered the more draconian treatment: The second edition in Cremona has the central leaf of chapter 8 sliced out, but the beginning and ending of the chapter were printed with uncensored material on the other side, so sheets of paper were pasted over this prohibited part. I'm not sure who would have been put off by the warning in Trigault's third edition, but it does offer a fascinating window on the mindset of the Vatican workers in the early seventeenth century.

Chapter 10

THE HUB OF THE UNIVERSE

THE PICTURESQUE Swiss town of Schaffhausen, still retaining its medieval center, sits in a leaf-shaped salient largely surrounded by Germany. In fact, this geographic oddity itself completely envelops several pockets of German territory, as I had learned in trying to take shortcuts in the vicinity. I had purchased an export car at the VW factory in Wolfsburg, but my tax-free status in Germany had expired, so unintentionally crossing into Germany was a considerable nuisance. But being in a salient north of the Rhine was more than a nuisance for the citizens of Schaffhausen in World War II. It was a disaster. On 1 April, 1944, a thousand Allied bombs mistakenly rained down on the precariously situated town.

On that spring evening a score of B24 bombers had headed toward the German chemical works in Ludwigshafen, twenty miles farther east at the tip of Lake Constance, but clouds and winds over France confused the formation, which lost its way. When the lights of a town appeared through a clearing in the clouds, crews of thirteen of the planes assumed they were over Germany and mistakenly dropped their deadly loads onto Schaffhausen. The damage and civilian casualties in neutral Switzerland prompted a diplomatic crisis for the United States. An apology and a million dollars in reparations were immediately forthcoming, supplemented by three million more in October.

Among the losses were nine paintings by the sixteenth-century Swiss artist Tobias Stimmer. When Stimmer decorated the famous astronomical clock in the Strasbourg cathedral, he included a portrait of Copernicus

*Wood-block portrait of Copernicus attributed to Tobias
Stimmer. The lily of the valley is the standard early
Renaissance icon for a medical doctor.*

(plate 2). As part of the painting he conspicuously inscribed the words *vera
efigies ex ipsius autographo depicta*—"a true likeness from his own self-portrait." Stimmer's caption has led scholars to conjecture that the lovely and
similar portrait of Copernicus now hanging in the city hall in Toruń is,
if not the actual self-portrait painted by Copernicus himself, at least a
copy of a sketch made by the multitalented astronomer. There is as well
a handsome woodcut by Stimmer, which, like the Strasbourg image,
shows Copernicus holding a lily of the valley, a Renaissance symbol for a
medical doctor.

Part of the American reparations went to establish the Tobias Stimmer Foundation, to help the museum in Schaffhausen acquire substitutes for their irreplaceable treasures. Among the pieces that survived the bombing is a curious sixteenth-century wooden instrument with a movable dial that can be used to calculate lunar phases. Some years ago the museum was able to purchase with the Stimmer Foundation funds the unique copy of the explanatory booklet that belongs with the instrument, and subsequently the foundation asked if I could provide a commentary to the book and instrument. I swiftly agreed, for there was a second reason I wished to visit Schaffhausen: The library next door to the museum (apparently undamaged in the raid) holds one of the most important copies of Copernicus' *De revolutionibus*. Thus, in a roundabout way, the American bombing of Schaffhausen in 1944 paid for my visit there in 1987.

The City Library sits in the charming medieval centrum of Schaffhausen. It is not large enough to support a rare book librarian. To see the Copernicus, I had to ask in the children's department, but I was allowed to take the book into the main reading room. There I opened it with some reverence. As an annotated copy, it is second only to the Reinhold volume that had precipitated the entire Copernicus chase, for it was thoroughly annotated by Michael Maestlin, a leading astronomer of the sixteenth and early seventeenth centuries, and Johannes Kepler's teacher. It was not the first time I had seen the book. I had examined and photographed it in 1972, and the library had provided a microfilm. The annotations were as fascinating as they were frustrating to read. The ink had partly faded, and still more troublesome, Maestlin wrote with an incredibly minuscule hand, so in many places, even using the microfilm, I could scarcely read what he had written. It required a firsthand examination to decipher some of the key marginalia.

Maestlin had been born in 1550, seven years after the publication of *De revolutionibus*, in Göppingen, a village about thirty miles east of Tübingen where he spent most of his professional career. As was the case with many young scholars including Kepler, his most famous student, he did his undergraduate studies at a preparatory school and came to the university to

M. MICHAEL MÆSTLIN.

ÆTATIS ≥ 8³⁹⁶ 1578

Michael Maestlin at age twenty-eight, from his Ephemerides novae *(Tübingen, 1580), author's collection.*

take his final exams and pick up his baccalaureate degree. Then he enrolled for the master's degree followed by the theological program, because Tübingen University's primary mission was to prepare young men for the Lutheran ministry. Thus educated, he was sent off in 1576 as an assistant pastor to Backnang, a post about twenty miles northwest of Göppingen. His time there was rather like that of young people today who serve in the Peace Corps before taking up their intended careers.

Maestlin's real talent lay in astronomy, and while a master's student he had invested in his copy of *De revolutionibus*. At age twenty-one, he had edited a new edition of Reinhold's *Prutenic Tables*, that set of numbers based on Copernicus to be used for computing planetary positions. While pastoring at Backnang, he published his observations of the Great Comet of 1577; like Tycho Brahe, he showed that the comet lay beyond the Moon, contrary to the accepted teachings of Aristotle. So it was only a matter of fulfilling his service appointment before he could move on, finally, at the age of thirty, to a mathematics professorship at Heidelberg, where he published the first of seven editions of his basic astronomy text-book. Four years later, in 1584, he received the call back to Tübingen.

As I sat with his *De revolutionibus* in the Schaffhausen library, I marveled at the multiple layers of annotations in the copy that demonstrated the lengthiest active use by a single owner. Inside the front cover was a hand-colored bookplate with a coat of arms, dating from 1584, when he had been recalled to Tübingen. But on the title page was an earlier inscription, "Ex libris M Michaelis Maestlini Goeppingensis Anno Domini 1570," and inside the back cover was a note saying that he had obtained the book from the widow of Victorin Strigel for one and a half florins. Strigel had been a student at Wittenberg, had founded the school that later became the University of Jena, and had gone on to become professor of theology at Leipzig, where he wrote an astronomy textbook. Elsewhere in the volume Maestlin's notes cite the third edition of *De revolutionibus*, which was published in 1617, and furthermore, the places censored or altered in 1620 by the Vatican Congregation of the Index were indicated with red emphasis marks. From 1570 to 1620—fifty years of annotating!

In the margins near the beginning of the book Maestlin penned a unique appreciation of what Copernicus had accomplished. No other copy of *De revolutionibus* contains anything comparable. Maestlin pointed out that "the arrangement presented in this book is the sort of structure in which all the sidereal motions and phenomena are explained very exactly. Therefore this hypothesis recommends itself to the intellect." Maestlin went on to comment that he thought many others would also agree with Copernicus' ideas if they hadn't been convinced long before that the Earth didn't move. Copernicus wasn't just playing a clever game, he wrote.

> The heavenly motions were at the point of collapse, and so he concluded that appropriate hypotheses were needed to explain these motions. When he noticed that the common hypotheses were insufficient, he eventually accepted the idea of the Earth's mobility, since indeed, it not only satisfied the phenomena very well but it didn't lead to anything absurd.

In fact, if anyone would straighten out the common hypotheses so that they would agree with the phenomena and allow no inconsistencies, then I would gratefully trust him; clearly he would bring very many to his views. But I see that some, even very outstanding mathematicians, have labored on this, yet, in the end, without results. Therefore, I think that unless the common hypotheses are reformed (a task that I am not up to because of my inadequate abilities), I will accept the hypotheses and opinion of Copernicus—after Ptolemy, the prince of all Astronomers.

Because in his elementary astronomy textbook Maestlin presented only the geocentric arrangement, his complete commitment to the Copernican system has always seemed somewhat problematic. The statement near the beginning of his copy of Copernicus' book would dispel any doubts except for his repeated use of the word *hypothesis*. Today when the word *hypothesis* is used to describe a scientific concept (such as evolution), many people tend to mentally add the word *mere*, to make a pejorative *mere hypothesis*. Sixteenth-century astronomers, working in a very different intellectual framework, used the word in a very different way. They generally viewed astronomy as a geometric (rather than physical) science, and the hypotheses were the geometric devices or arrangements used to explain celestial motions. Later on Maestlin would scold his student Kepler for dragging physics into astronomy, which he believed was inappropriate. So Maestlin may well have harbored reservations about the physical reality of the Copernican arrangement even while accepting that it was the best explanation for the planetary phenomena.

Toward the back of the book long notes in Maestlin's microscopic hand revealed that he tried to pursue a fiercely technical question that Copernicus had left hanging and which could potentially serve as another convincing (but subtle) argument for the heliocentric arrangement. Copernicus had referred all of the planetary motions not to a fixed Sun but to the center of the Earth's orbit. This would have made little difference except that Copernicus had noticed that the distance between the Sun and the center of the Earth's orbit had diminished since Ptolemy's day. If the center of the Earth's orbit were the true center of the universe, then it wouldn't make any difference with the planets' orbits if the Sun bobbled back and forth with respect to Earth's orbit. On the other hand, if the Sun was the truly fixed reference point for the planets' orbits, then their own centerings would change over time as seen from the bobbling center of the Earth's orbit. So Maestlin set out to see if he could establish this effect. Had he been able to find it, there would have been one further connection between the Sun and the planets, a fine argument for the centrality of the Sun in understanding the planetary arrangement. But he failed, not for any shortcoming with his mathematical technique but because Ptolemy's observations were not accurate enough for the required comparison.

Maestlin's annotations, like Reinhold's, were thick in the technical, later sections of *De revolutionibus*. But unlike Reinhold, Maestlin also had some very cogent remarks about the front matter of the book; in fact, they are the most fascinating front-end comments in any extant copy. At the very beginning, above the anonymous introduction to the reader, the multiple layering of notes exhibits itself. Maestlin started off by saying, "This preface was added by someone, whoever its author may be, (for indeed, its weakness of style and choice of words reveal that it is not by Copernicus)."*

On the top margin of the facing page Maestlin added another note.

NB: Concerning this letter, I found the following words written somewhere among the books of Philipp Apian (which I bought from his widow); although no author was given I could recognize Apian's hand:

On account of this letter Georg Joachim Rheticus, the Leipzig professor and disciple of Copernicus, became involved in a very bitter wrangle with the printer, who asserted that it had been turned over to him with the rest

---

*Ever since reading that, I've wished I could read Latin well enough to make such a judgment. Considering that its author misled a great many readers into supposing that the introduction was by Copernicus himself, I have to assume that it takes a particularly astute and perceptive critic to detect such nuances.

of the work. Rheticus, however, suspected that Osiander [the proofreader of Copernicus' book] had prefaced it to the work. If he knew this for certain, he declared, he would handle that fellow so that in the future he would mind his own business and not slander astronomers any more. Nevertheless [Peter] Apian told me that Osiander had openly admitted to him that he had added this all by himself.

It took me some later detective work to sort out the origin of this confusing note—confusing because more than one Apian is mentioned. I realized that both Philipp Apian and his more famous father, Peter Apian, were involved in Maestlin's note after I discovered another copy of the quoted passage in Munich. The *De revolutionibus* in the Bavarian capital had been there since 1571, whereas the copy of the note that Maestlin had seen (not necessarily in a copy of Copernicus' book) was in his hands after 1589, when Philipp Apian had died. So here is how I reconstruct the story. Peter Apian, a well-known author and astronomy professor at Ingolstadt north of Munich, got the information about the anonymous introduction straight from Osiander himself and mentioned it to a colleague, who noted it in his copy of *De revolutionibus*. At some point the young Philipp Apian saw the note, probably after his father had died in 1552, and he made a verbatim copy of it. The original *De revolutionibus* containing the note was acquired by the banker Johann Jacob Fugger, who had it rebound (thereby trimming off part of the note). In 1571, when he was bankrupted by his passion for books, Fugger sold the *De revolutionibus* to Duke Albrecht V of Bavaria, founder of the library in Munich, where it has been ever since. Philipp Apian's copy has disappeared, but what is probably the original, as well as Maestlin's third-hand transcription, both survive.

This was not the end of the Osiander story, however, for Maestlin wrote a third, terse comment above the anonymous introduction: "NB: I know for sure that the author of this letter was Andreas Osiander (plate 7f)." What made him so sure? The answer revolves around Maestlin's most famous student, the primary reason that Maestlin himself is remembered today.

LIKE MAESTLIN, Johannes Kepler was born near Tübingen, studied in a preparatory school before going to the university to study for his master's degree, and then went into the theological program, fully expecting to become a Lutheran clergyman. When, instead, he was sent out to be a mathematics and astronomy teacher, he complained that nothing had indicated that he had a special talent for astronomy. In evidence, he was a straight A student *except* in astronomy, where he got an A–.* Nevertheless, astronomy was in his background. Kepler recalled that, when he was six years old, his mother had shown him the Comet of 1577. Also, he may very well have inherited his *De revolutionibus* through his family, possibly from a Nuremberg bookseller named Kepner, who may have been an ancestor.

I remember vividly the circumstances when I first saw Kepler's copy in Leipzig in 1972, the same field trip that had originally taken us to Schaffhausen. It was the first time Miriam and I had penetrated deep into East Germany beyond East Berlin. The country was a drab police state, but with scattered friendly though apprehensive persons cautiously willing to make contact with the outside world. Since we had got behind the iron curtain, I was keen to see the Wittenberg archives and to find any traces of Erasmus Reinhold, which was what made the trip particularly memorable. But first we had to attend to Leipzig.

Edition Leipzig was a major East German publishing house eager to bring in hard currency through splendidly printed art books and reproductions of library treasures. One of the great typographical triumphs of the sixteenth century, printed just three years before *De revolutionibus*, was Peter Apian's *Astronomicum Caesareum*—literally, "Astronomy for the Emperor," and the princely book Tycho had glowingly inscribed to Wittich. It was a giant folio with astronomical diagrams full of rotating

---

*In the Tübingen grade reports, a capital *A* was an A, and a lowercase *a* was an A–.

parts, brilliantly hand colored. The most complex set of volvelles, seven layers deep, served as an analog computer to simulate the Ptolemaic epicyclic theory for finding the longitude of the planet Mercury. Edition Leipzig used a disassembled copy from the library in Gotha to make a spectacular facsimile. Unfortunately, while its facsimile was a typographical tour de force, the actual assembly of the moving parts was thoroughly botched, with some volvelles on the wrong pages and others pasted down so they wouldn't turn properly. On the pages of the *Journal for the History of Astronomy* I had called attention to this misassembly, and subsequently Edition Leipzig had invited me for a consultation.

I explained to my hosts that in addition to trying to figure out what to do about the faulty facsimile,* I hoped to explore the Wittenberg University archive. In response they broke the news that the famous old University of Wittenberg no longer existed. It had long since been combined with the university in Halle. And there was another problem. As the editors pointed out, our East German visa was good only for the Leipzig circle, and if we drove over to Halle, we would be dangerously conspicuous. They offered instead to send us over to Halle the following day by train with one of their assistants, thinking that no one would then notice.

The assistant was enormously pleased to accompany us to Halle. On that particular day the radical American activist Angela Davis was in Leipzig for a rally, and each company had a quota of employees obliged to turn out for the demonstration. The young assistant who accompanied us to Halle had been assigned as a "volunteer" to attend the rally, something she did not relish, so she was delighted to be escorting us instead. We were surprised at how freely she made her sentiments known, in English, on the train.

In Halle we got a warm welcome. The university library's copy of

---

*Edition Leipzig agreed to print a repair kit for the *Astronomicum Caesareum*, but abandoned the project when so few buyers caught on that something of this sort was needed. Since then I have used the color proofs of its aborted repair project to correct more than a dozen copies—typically requiring nearly eight hours of work on each one—and I have distributed repair kits to about a dozen other owners.

Copernicus' book had been missing for some time, but the librarian was eager to show us some of their other rarities. The crown jewel was the dean's book from sixteenth-century Wittenberg. I immediately recognized the clear, neat hand of Erasmus Reinhold. The deanship had been a bureaucratic office passed quickly around from one teacher to another, but because Reinhold's handwriting was so legible, some of the other deans asked him to do the honors. A pair of special lectures recorded in Reinhold's hand jumped out, one by Reinhold himself on astronomical hypotheses, and the other "Against the Anabaptists." Since I come from a long line of Anabaptists, this topic particularly resonated with me. I had brought along my Nikon, but I had only color film. This was the only time I've made a microfilm on Ektachrome.

Back in Leipzig I naturally went to see both the *Astronomicum Caesareum* and the first-edition *De revolutionibus* in the university library. I was especially thrilled to see Kepler's Copernicus, but in retrospect I missed several of the most important points. Since Edition Leipzig had already issued a facsimile of the book, I wasn't surprised to see on the flyleaf Kepler's Latin translation of a long Greek poem by the Leipzig humanist Joachim Camerarius, nor did I have a sudden burst of insight when I saw Osiander's name written above the anonymous introduction, the *Ad lectorem*. What Edition Leipzig had *not* included in its facsimile, quite rightly as it turned out, was an old inscription concerning the earliest copy of Copernicus' book acquired by the university. It had been clipped out of the original copy when it was sent out for auction as a duplicate, and hence it was an important document for the history of *De revolutionibus* in general, and so I was keen to see it even though it was totally irrelevant to Kepler's copy in particular.

What I failed to photograph were two annotations that attracted my attention only after I began to understand the layering of the annotations in the book. Most, but not all, of the marginal annotations are in Kepler's hand. Many small interlinear corrections are in the hand of its original owner, Jerome Schreiber, who had received the book as a gift from the printer, as he noted in a corner of the title page. Schreiber was from Nuremberg but studied at Wittenberg and was for a while a mathematics

1a. The "Eames machine" in the IBM Copernican exhibit demonstrated the equivalence between the Ptolemaic epicyclic model (left) and the Copernican orbits for Mars—the rods remained parallel as the circles rotated in each system.

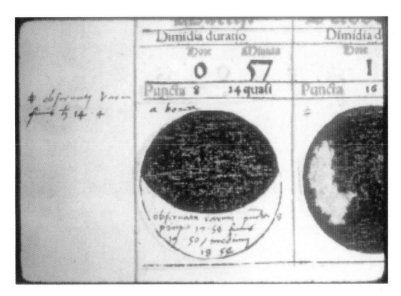

1b. Copernicus' eclipse annotations in his copy of Johann Stoeffler's Calendarium Romanum magnum.

NICOLAICO
PERNICIVE
RA·EFIGIES
EX·IPSIVS
AVTOGRA
PHO·DEPI
CTA·

2. *Tobias Stimmer's portrait of Copernicus, part of the decoration of the great astronomical clock in the Strasbourg Cathedral.*

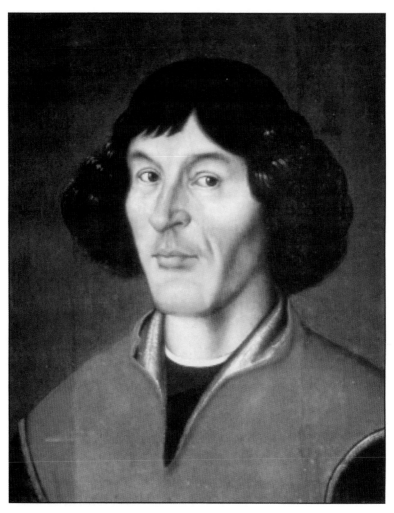

3. Nicolaus Copernicus, Toruń town hall, presumably based on the self-portrait mentioned by Stimmer (see facing page).

4a. *The three volumes given to Copernicus, with Rheticus' presentation inscription.*

4b. *Erasmus Reinhold's summary on the title page of his richly annotated* De revolutionibus.

5a. The Copernican Library preserved at Uppsala University. Copernicus' copy of the Regiomontanus Epitome of the Almagest *has not been located, and another copy (foreground) was substitued for the picture.*

5b. Charles Eames photographing the Copernican books in the Uppsala University Library.

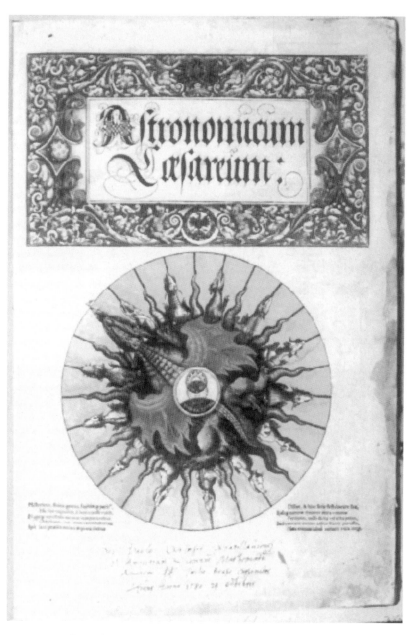

6. Peter Apian's **Astronomicum Caesareum** *with Tycho Brahe's presentation inscription to Paul Wittich.*

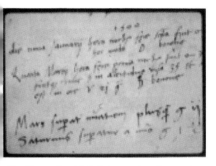

7a. Copernicus' brief observational notes in the so-called "Uppsala notebook."

7b. Thomas Digges's signed endorsement of Copernicus: "The common opinion errs."

7c. Herwart von Hohenburg's colorful annotations in his De revolutionibus.

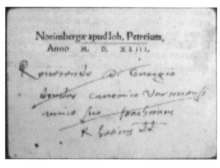

7d. Rheticus' dedication inscription to the Varmian canon George Donner.

7e. Wilhelm Schickard's artistic talents are evident in his marginal drawing of a triquetrum, one of the few instruments mentioned by Copernicus.

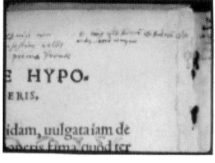

7f. Michael Maestlin's note identifying Andreas Osiander as the anonymous author of the Ad lectorem on the back of the title page of Copernicus' book.

8a. Owen Gingerich with Shida Weng and Peishan Li examining the second and third editions of De revolutionibus *and Kepler's* Rudolphine Tables *in the rare book room of the National Library of China, Beijing, 1985.*

8b. Owen Gingerich examining the most important first edition of De revolutionibus *in the Western Hemisphere, in Yale University's Beinecke Library. His own copy of the second edition lies just beyond the first.*

teacher there. He was an insider, so to say. That turned out to be quite important, because when I eventually examined those corrections more carefully in my copy of the facsimile, I discovered something curious. Whereas the corrections in the first three-quarters of the book came directly from the printed errata leaf that accompanied some copies, the corrections continued to the very end, well beyond where the errata leaf left off. After I discovered this same extended errata set in a few more copies, I realized that the insiders had access to a more complete list than the printer Petreius had supplied to his customers.

After I saw the extended errata marked in several copies, I caught on that a marginal note on folio 96 was connected with these corrections, although it was simply a comment and not a correction. At that point in the text Copernicus was considering what was the true fixed center of the cosmos. Was it the Sun itself, or was it the center of the Earth's orbit? Because Copernicus believed that the stars were in a distant spherical shell, the question was whether the Earth's annual round, the "great orbit" as Copernicus put it, was neatly centered with respect to the stars (thereby putting the Sun slightly off-center), or whether the Sun itself was dead centered so that the Earth was closer to the starry shell in January than in June. If the Sun was the hub of the universe, Copernicus truly had a heliocentric system. If the hub was the center of the Earth's orbit, eccentrically offset from the nearby fixed Sun, it was a heliostatic system. This was an unresolved mystery in the book, for Copernicus hedged on the issue. However, the marginal note in a few of the copies indicated that Rheticus' *Narratio prima* said more. The one in Schreiber's hand reads (in Latin, of course), "These are in the *Narratio* of Joachim [Rheticus]. For in this work they are omitted." In fact, the *Narratio prima* did not discuss the question explicitly but simply assumed throughout that the Sun itself was the center of the universe.

What follows is an exercise in minutiae, but one that ultimately offers a most intriguing insight into Kepler's student-teacher relationship with Michael Maestlin. Below Schreiber's note is another, looking at first glance very much like Kepler's hand, yet clearly distinct from Kepler's annotations elsewhere in the book. In fact, I believe it matches Michael

Maestlin's hand more closely than Kepler's.* It is not the minuscule writing that characterizes the notes in Maestlin's own *De revolutionibus*, but very much like the handwriting in his letters to Kepler. The note states, "What can be accepted about this question is that from Book V they make either the sun fixed or else the centers of all the planets slightly displaced with it." Maestlin not only wrote a very similar comment at the same place in his Schaffhausen copy but also marked places in Book V corresponding to his statement in Kepler's copy.

Why do I get excited about something as esoteric as this? Because the presence of this little note tells us that Kepler showed his copy to his teacher, and that's why Maestlin was so sure it was Osiander who had written the anonymous introduction to Copernicus' book. There it was in Kepler's copy, in black and white, coming straight from Schreiber, a Wittenberg insider. More than that, Maestlin's little annotation suggests that he and Kepler specifically discussed this point, what was the hub of the universe. And on the day when Kepler went to work for Tycho Brahe in 1600, his notebook shows that the very first task in his research program was to adjust the orbit of Mars so that it referred to the Sun rather than to the empty point that happened to be the center of the Earth's orbit. This became an essential part of Kepler's physical approach, and a fundamental principle that led to his successful reworking of the heliocentric details. It was presaged by that fateful student-teacher conversation at Tübingen.

Kepler is most famous for discovering the elliptical shape of the planetary orbits, and so another of the marginal comments also seems highly relevant. On folio 143 there appears the single Greek word ελλειψις—

---

*In fairness to full disclosure, I have to say that two of Germany's leading experts on Kepler's hand are fully convinced that I'm wrong, but their opinion does not come to terms with the fact that such a similar annotation also appears in Maestlin's *De revolutionibus*. Clearly there is a close connection between the two notes, which both begin with the identical words *Quae de hac quaestione . . . possunt*; it would be exceedingly odd if Kepler copied just part of the annotations and nothing else from his teacher's book. The most distinctive handwriting feature of the short note is the way the tall *s* and *t* are joined in the word *quaestione*. I searched many pages of Kepler's manuscripts and found that he used such a combination only very occasionally. For Maestlin the conjoined letters are frequent, including in his own name, with a closely matching appearance.

*Marginal annotations by Jerome Schreiber and Michael Maestlin
(folio 96, left) and Johannes Kepler (folio 143, right).*

that is, *ellipse*—together with the same sort of emphasis marks that Schreiber used to highlight the passage on folio 96.

When I first saw that book in Leipzig, I assumed that it was Kepler who had written ελλειψις in the margin, and I hadn't made a color slide of it. Later, when I had discovered more information about the double layer of annotations and the evidence that it was likely Schreiber's handiwork, I had to worry about which one wrote it. Handwriting comparison didn't help because it was in Greek letters, not Latin. I decided I had better go back to Leipzig for a close scrutiny of the ink, only to learn that the book was away on exhibition. Eventually I obtained excellent color transparencies, which left no doubt that it was indeed Schreiber's ink in the book Kepler had inherited.

Did this single, crucial word give Kepler the clue for his greatest discovery? And was Copernicus himself on the scent of the planetary ellipses?

The fact that we can definitively answer the second question with a resounding "No!" comes as a considerable surprise to many of my colleagues.

IN 1985 A Louisville Seminary theologian named Harold Nebelsick published a fascinating but wrongheaded book entitled *Circles of God*. He made the provocative claim that the requirement of using circles and only circles to explain celestial motions was a theological invention of the ancient Greeks and a bad idea that had held astronomy in thrall for two millennia. Included in his contention was the insinuation that Copernicus had failed because he stuck with circles, not noticing that the orbits of the planets were really ellipses. But there was no way Copernicus could have found the ellipse, because the observations he had weren't nearly accurate enough.

That there is a problem becomes very clear if you draw an ellipse corresponding to the orbit of Mars. Take an ordinary piece of letter paper, put two tacks in about an inch and a half apart to represent the foci of the ellipse, then loop a string around them to guide the pencil using the page to its fullest, and you will get an ellipse. If you then draw the corresponding circle (with a radius of nearly four inches), you will be hard put to tell which is the ellipse and which is the circle, because the difference is about the width of the pencil line. Naturally astronomy textbooks don't show it this way, because they can't make the point about ellipses unless they enormously exaggerate the eccentricity of the ellipse. So for centuries, beginning with Kepler himself, a false impression has been created about the elliptical shape of planetary orbits. The eccentricity of planetary orbits (that is, their off-centeredness) is quite noticeable—even Ptolemy had to cope with that—but the ellipticity (the degree the figure bows in at the sides) is very subtle indeed. Observations of Mars must be accurate to a few minutes of arc for this tiny ellipticity to reveal itself. This is near to the limit of naked-eye acuity, and such observations simply weren't available to Copernicus or any of his predecessors. Not until Tycho Brahe's massive and precise observational campaign was the requisite data bank available, and within fifteen years of that availability, Kepler, with Tycho's record books in hand, found the elliptical compression of the orbit.

What about that passage in *De revolutionibus,* the place where Schreiber had written *ellipsis* in the margin? Could Copernicus have been on the trail of the ellipse? The Polish astronomer realized that when he replaced the Ptolemaic equant with a small epicyclet, the resulting path would not be exactly circular. In fact, though he never said so explicitly, the combination of deferent and epicyclet produced an ellipse. But it's a wrong ellipse, one that bowed out where the correct ellipse bows in. Copernicus' curve was an artifact of his model and had nothing to do with the true trajectory of the planet. Still, it's very romantic to speculate that the Greek word Schreiber penned in the margin had some subliminal power of suggestion on Kepler.

Copernicus, like Ptolemy centuries earlier, used very few observations to establish the parameters of the planet's orbit. He was no doubt blissfully unaware that for a brief period every seventeen years both his and Ptolemy's predictions for Mars went horribly wrong. For Ptolemy, Mars lagged behind the predictions by five degrees; for Copernicus, the ruddy planet went ahead by about four degrees. Apparently Kepler was the first person to comment on this, and he probably noticed it only after he had corrected the orbit of Mars for other reasons.

When I first became aware of this anomaly, I assumed it was caused by some erroneously chosen constant that entered into the calculations. Eventually I tried tweaking the numbers used in the model. The results proved very disconcerting. While I could make the error go away in one place, it always popped up somewhere else. Clearly there was a fundamental defect in the model itself, and it couldn't just be the lack of an ellipse.

Copernicus failed in this matter not because he hadn't caught on about the ellipse but because he wasn't Copernican enough. It was Kepler who remarked, in a slightly different context, that Copernicus was unaware of his own riches. If Copernicus had really believed that the Earth was just one of several planets, he should have treated them all the same. That would have been the "Copernican" thing to do. But Ptolemy hadn't used an equant for the Sun, and therefore Copernicus didn't use his equant substitute for the Earth. (The Earth and the Sun are the two celestial objects at opposite ends of the connecting line, and the mathematics works the same way regardless of which end of the stick is considered the stationary reference point.) The bottom line: In the Ptolemaic system the Sun moved around its circle at a constant speed—it just looked as if it moved at different speeds because that circle was eccentric to the earth. Likewise in the Copernican system the Earth moved around its circle at a constant speed—the Sun just looked as if it moved at different speeds because it wasn't at the center of the Earth's circle.

And this, Kepler believed, had to be wrong. If Mercury, the planet closest to the Sun, moved fastest, and Saturn, the most distant planet, moved slowest, then this was because Mercury, being closer, soaked up more of the Sun's motive power and thus naturally moved faster. But in winter the Earth was closer to the Sun than in summer, and Kepler reasoned that it should actually be going faster in its orbit in winter. That was physics, and Kepler, as the world's first astro-physicist, worked out the consequences. Maestlin rapped his student on the knuckles for that. He wrote to Kepler, "I think that one should leave physical causes out of the account, and should explain astronomical matters only according to the astronomical method with the aid of astronomical, not physical, causes and hypotheses. That is, the calculation demands astronomical bases in the field of geometry and arithmetic."

But Kepler persisted. He had to adjust the position of the Earth's orbit to make it work, and when he did, the periodic five-degree error in the Mars predictions just melted away. That was the biggest single correction that Kepler made in predicting the positions of the planets, and he doesn't get much credit for it because the astronomers who later selected three of Kepler's discoveries and numbered them as three laws (perhaps to match Newton's three laws) simply passed over this one as being too obvious.

Even before he made this discovery, Kepler found a tricky way to calculate, quite accurately, the longitude of Mars as it went in orbit around the Sun. But when he tried to locate Mars *as seen from the Earth,* he ran into trouble. The calculation that worked so well in tracking the east-west motion of Mars around the sky simply wouldn't work for Mars's north-south deviations in latitude. When he fixed that, he ended up with a maximum error of around half a degree. This was already ten times better than Ptolemy or Copernicus had achieved, but it wasn't good enough for Kepler because it didn't match Tycho's excellent observations. He could have used a jury-rigged, physically inconsistent scheme to get the longitudes almost five times better (or fifty times better than Copernicus' maximum error), but to Kepler that lacked reality because it didn't give correct latitudes, and unlike his teacher, he was a thoroughgoing realist. Kepler tried an ellipse, not quite the right one, as an approximating curve. And then came a moment of truth. "Oh ridiculous me!" he wrote. "I could not find out why the planet would rather go on an elliptical orbit. . . . With reasons agreeing with experience, there is no figure left for the orbit of the planet except a perfect ellipse."

Had he got the clue from that little marginal note in his *De revolutionibus*? I doubt it, but who knows what pathway triggered his imagination?

The ellipse would have been hard for Copernicus to accept because he was so thoroughly committed to the principle that celestial motions should be explained in terms of uniform, circular motion, but in the end he surely would have approved the quest for a physically real system.

Chapter 11

THE INVISIBLE COLLEGE

IT WAS DUMB luck that Miriam and I missed the flight out of Oklahoma City on a Sunday in February 1993. The airline counter was suspiciously empty, and we should have had an instant foreboding that something was wrong. "You're quite early for the flight," the attendant cheerfully informed us. But then she examined our tickets. "Woops—your plane left half an hour ago! But don't worry. You can get to Boston via Chicago instead of Dallas, and neither segment is full."

The nature of my blunder dawned on me after a few moments. At that hour our connection in Dallas was boarding, not the link from Oklahoma City, which had long since departed. We had showed up at the right time, but in the wrong state! But if I hadn't made that stupid mistake, I might never have learned where Jofrancus Offusius was born nor unscrambled his connection with *De revolutionibus*, and I might even still think that his name was an erudite Greek pseudonym based on the celestial constellation Ophiuchus, the Serpent Bearer.

We had come to Oklahoma to attend a conference at the university in Norman, and while there I intended to take advantage of its outstanding collection of rare books in the history of science. I had inspected its copy of Copernicus' book early on in my search, in fact, so near the beginning that I hadn't been very savvy about what to look for in the annotations.

My colleague Bob Westman had come to the conference too, and when I learned that after the conference the library was being opened specially for him Sunday morning, I decided to come along to have a fresh look at the *De revolutionibus*. Had I remembered when our plane was *really* scheduled to leave, I never would have taken the extra time to sit with the Copernicus book again, and to copy out some of the more interesting notes. And then I would have overlooked an important clue to a puzzle about the Copernicus books that had been baffling me for nearly a decade.

Unfortunately, not every owner of *De revolutionibus* bothered to put his name in it, and in particular I had a cluster of copies with very similar annotations but no clearly defined original annotator. One potentially useful feature of Copernicus' book, for those readers who actually wished to calculate the position of a planet, was the so-called mean motion tables, the first step in locating a planet. But to use the tables, it was necessary to have a starting position, or *radix*, something that Copernicus unfortunately buried in the text where it was a nuisance to find, so this group of annotators had written in a starting position for 1550 at the bottom of each relevant table. Thus I called whoever started the sequence the "Master of the 1550 Radices," but his mysterious identity had eluded me.

As I sat with the first-edition *De revolutionibus* at the University of Oklahoma that Sunday morning, one unusual detail caught my eye: three marginal references that attributed the original annotations to someone named Vesalius. Westman allowed that he had noticed the name earlier, but like me, the only Vesalius he had ever heard of was the famous medical doctor whose *De humani corporis fabrica*, published in 1543, the same year as *De revolutionibus*, had revolutionized the study of human anatomy. Did Andreas Vesalius also have a secret life as an astronomer? Strange things had turned up in the census, but nothing quite as astonishing as that.

Thanks to the helpful clerk at American Airlines, we made it back to Cambridge only an hour after our originally scheduled plane. A few days later, in reviewing my notes, I suddenly noticed that the longest note I had transcribed, on folio 127, matched an annotation that had previously turned up in seven other copies of the book. What was really maddening was that each of the notes was written in the first person—"*Ego reperi* . . . (I have found . . .)—and yet they were written in eight different hands.

A seemingly valuable clue to the elusive original annotator had come not from one of the books in the Master of the 1550 Radices series but from a quite different copy of *De revolutionibus*. It was in Yale's Beinecke Library, the copy that had exhilarated me so much years before when I had identified it as a long-lost one that had been owned by someone with at least an indirect connection to Paul Wittich. At the back of the Yale book an early owner had written that Thaddeus Hagecius—the personal physician to Emperor Rudolf II and a sometime astronomical author himself—had found out from Paul Wittich about three errors in Copernicus' book. These he listed in their stark triviality. The last two, minor arithmetic mistakes in the great cosmological chapter in Book I of *De revolutionibus*, were pretty inconsequential. But the first related to folio 127, the very place where the Master of the 1550 Radices copies had such a huge annotation.

In at least one of the eight copies, in a *De revolutionibus* in Debrecen, Hungary, the same three errors were marked, in so precise a way that it couldn't be coincidence.* A slender clue indeed, but to me, it smelled of Central Europe and in particular of Georg Joachim Rheticus, the rebellious young man who had been Copernicus' only disciple and who had taken *De revolutionibus* to the press. Subsequently, he had been persuaded by a particularly high salary offer to leave Wittenberg for Leipzig, and had eventually settled in Cracow, from where he had later connections with both Hagecius and Paul Wittich's family.

But why Rheticus? He had been the professor of mathematics at Wittenberg when Erasmus Reinhold was the professor of astronomy there. The census had turned up more than a dozen books with partial copies of Reinhold's annotations, but from Rheticus there were only a couple of presentation copies without any significant annotations. Surely the young man who made the trip to Poland, who first learned of the heliocentric system, and who brought the manuscript back to Germany for publication

---

*Whether any of the other copies included these three errors was hard to know, since I didn't have microfilms of all of them, and in the first instance of examining the books these errors would have escaped notice as being so seemingly innocuous. As in most good mystery stories, minor details assume great importance only in retrospect. It turned out that a copy in Edinburgh also had these errors marked—even more thoroughly than the copy in Debrecen.

would have taught the details to a generation of students. Where was his teaching tradition? It was curiously absent from the annotated copies.

Since finding Rheticus' annotations would be a major coup, I had a secret hope that these eight books might hold a precious record of his own master notes. Rheticus had carefully computed an ephemeris for 1551, that is, an almanac giving planetary positions for every day of the year, of course based on the tables in *De revolutionibus*. If anyone aimed to embark on such a task, probably the first step would be to calculate the mean position of each planet for some suitable starting date, such as 1 January 1550, and write that in a convenient place. And that is precisely what I had found in the tables of seven of these eight books, which helped me recognize their affiliation in the first place. Only the Oklahoma copy lacked these numbers, and that is why I hadn't caught on right away that it belonged to the set.

Was the mysterious Master of the 1550 Radices, in fact, Rheticus?

In 1992 my Polish colleague Jerzy Dobrzycki came to Cambridge for a research visit, and I put the suggestion to him. He lost little time in shooting it down. Though I had been collecting the materials and worrying about the problem for some years, I had not yet made a definitive transcription and translation of the longest marginalium, the one on folio127. Jerzy set to work on my preliminary notes and microfilms. "This is heavy criticism of what the annotator thinks are errors in this very technical passage," he pointed out, "but look here at the end. He says maybe Copernicus didn't do it but entrusted this part to a student. Since Rheticus was Copernicus' only student, he would hardly have written that."

So it was back to the drawing board to find another candidate for the Master of the 1550 Radices. One of the eight copies, in the Bibliothèque Nationale in Paris, actually had inscribed on its title page the name of the student who had copied the notes: Jean Pierre de Mesmes, sometime astronomer of Paris, about whom very little is known except that he put out an elegant but rather derivative book entitled *Les institutions astronomiques*. In one marginal note, not found in any of the other copies of this family of annotations, he had added a value for the precessional motion "from my teacher Jofrancus for this current year 1557." And, near the end of the book, he wrote "Johannes Franciscus,* not your ordinary astronomer, made a wonderful Copernican instrument for Master Rousseau."

Jofrancus, not a very common name, clearly referred to one Jofrancus Offusius, who in 1557 published in Paris an ephemeris for that year. Therein he referred to himself as a German, but he gave very little hint as to his origins, position, or patrons (if any). Not only did de Mesmes add to his own book those few notes about Offusius, but he also changed a few of the first-person "ego" notes to refer to Jofrancus. However, he left others as "ego," which made the situation very murky. And the fact that the Oklahoma copy changed three of the other "ego" notes to "Vesalius" muddied the waters still further. Added to this was yet another curious detail: A close copy of the de Mesmes version, a book also now in Paris, bore a note on its title page saying "from the hand of C. Peucer." (Caspar Peucer was Reinhold's successor in Wittenberg.) We could trace its route from Germany to present-day Paris, but it did seem preposterous to suppose that it had been annotated in Catholic Paris, found its way to Lutheran Wittenberg, and then back to prerevolutionary France.

Gradually, despite the confusion, it became obvious that Jofrancus Offusius had to be a principal in the case, but it wasn't clear whether he alone had written the basic set of notes. Was a colleague named Vesalius also in on the study?

To crack this aspect of the case required a map of the lower Rhine Valley and two more pieces of information: a letter that Offusius wrote in 1553 to the Elizabethan mathematician and spiritualist John Dee, in which he signed his name as J. F. van Offhuysen, and a reference to him as Johannes Franciscus Geldrensis by the Italian mathematician, astrologer, and medical doctor Girolamo Cardano. Geldrensis was Offusius' toponymn, that is, his geographic name, meaning that he came from the town of Geldern.† Today the town of Geldern sits near the Rhine just across the Dutch border in Germany. Nearby, on the Rhine itself, is

---

*Johannes Franciscus is simply a longer form of Jofrancus.
†The use of a geographic identifier was very popular in sixteenth-century central Europe, and nearly all the students matriculating at Wittenberg or Leipzig signed in with one.

Wesel, whence the toponymn Vesalius. And not far upstream is Oberhausen (or Overhausen, as it is spelled on the map of Westphalia in Gerardus Mercator's atlas of 1568), which conceivably relates to the dialect family name Offhuysen or Offusius. Dobrzycki had come back again in the summer of 1993 to work with me on the census, and there was great glee on his countenance when he held the *National Geographic Atlas* in his hands and convinced me that Vesalius and Offusius had to be one and the same. The Master of the 1550 Radices was surely Jofrancus Offusius.

We easily learned that Offusius had published one book in his lifetime, *Ephemerides* for 1557. I was a little proud of myself for having judged correctly the significance of those radices, for he surely added those base positions for 1550 to the tables of his *De revolutionibus* to simplify his calculation of the ephemerides. In our sleuthing we also discovered that Offusius had spent 1552 in England, with John Dee as his patron. Much later, in 1570, Offusius' widow issued a posthumous astrological publication by her late husband, *De divina astrorum facultate* (On the Prophetic Power of the Stars). This latter volume greatly annoyed the aging, crotchety Dee, who cried plagiarism for some of his astrological aphorisms, but modern commentators have given Offusius the benefit of the doubt.

One final but important point still had to be sorted out. Was the original Offusius copy of *De revolutionibus* extant? As the evidence began to point ever more strongly to Offusius as the perpetrator, I argued that the copy now in Debrecen, Hungary, was the likeliest candidate. This copy had an interesting pedigree, having been owned by the Hungarian humanist Johannes Sambucus, who had made a book-buying trip to Paris in 1559 and another in 1561–62. Since this is around the time when Offusius mysteriously disappeared from the Parisian scene (as mysteriously as he had suddenly emerged there several years earlier), the book could have been sold from his estate around that time. Besides, the marginalia had the kind of cross-outs that might be expected from the annotator himself.

All the other seven copies of the Master of the 1550 Radices series were demonstrably derivative, except for the one at the National Library of Scotland. How the Edinburgh copy got there isn't known. But what distinguished it from all the others was the presence of about fifty pages

Owen Gingerich reading microfilms with Jerzy Dobrzycki, in Gingerich's Harvard-Smithsonian office in 1979.

of astronomical notes bound at the end of the volume. These notes were in the same hand as the annotations in the margins, and by their nature they were clearly holographs—that is, written in the hand of the author himself. Recognizing that they were, in fact, Offusius' originals was not like a bolt from the blue, a sudden denouement when the detective assembles all the suspects and announces the inescapable line of logic that led to the inevitable solution. This conclusion was gradual, increasingly convincing, but ultimately irresistible, and very, very satisfying. For years the unresolved issue—who was the mysterious annotator of this major cluster of copies?—had stood as an obstacle in the path of publication of the Copernican census. Many misleading clues and false pathways had prolonged the research, but at last a definitive solution was in hand. The Edinburgh copy, not the one in Debrecen, had to be Jofrancus Offusius' original working text.

WITH THE mysterious annotator unmasked, Dobrzycki and I sat back and considered whether we had merely solved another *New York Times*

Sunday crossword puzzle, or if we had really learned anything interesting about Renaissance astronomy.

One fascinating insight from the Offusius case came from the opportunity to see his reaction to the heliocentric cosmology, as documented in the astronomical notes at the end of his copy of Copernicus' book, now in Edinburgh. There he cautiously praised Copernicus' accomplishment and criticized the timid instrumentalist position of the anonymous advice to the reader (*Ad lectorem*) that had been surreptitiously added by Andreas Osiander, but Offusius stopped short of a thorough endorsement of heliocentrism. Such matters, "when the arguments from geometry and physics are inadequate," must finally be settled by the Bible, he declared (as Osiander had demanded), and he went on to cite a series of Old Testament texts in favor of a fixed Earth. There was Psalm 19, "He set the tabernacle for the Sun, which is as a bridegroom coming out of his chamber, and rejoiceth as a strong man to run a race. His going forth is from the end of the heaven, and his circuit unto the ends of it," and Psalm 93, "Thou hast fixed the Earth immovable and firm, thy throne firm from of old," and the beginning of Ecclesiastes, "But the Earth abideth forever. The Sun also ariseth, and the Sun goes down, and hastens to his place where he arose."

In sixteenth-century Europe the Bible had a unique status among books. It was universally accepted that the Bible had literally been dictated by God. Though Offusius was to some extent open-minded about the Copernican cosmology, in the end he fell back on the Bible, which makes all the more remarkable the positions of Copernicus and Kepler and Galileo, each of whom was deeply religious.

BEFORE WE zeroed in on Offusius and his group, no one had any notion about a serious technical study of Copernicus' book in Paris in the sixteenth century, or for that matter, in any primarily Catholic setting. Yet here was an astrologer and mathematical astronomer, not a professor at the university but nevertheless with his own atelier of students eagerly copying out his insights into Copernicus' text. Curiously enough, none of Offusius' copyists had discovered that their master stumbled in his long critique on folio 127. He erred in his recalculation of the initial position of the Sun and Moon. But it was, after all, the most complex spot in Copernicus' entire book, a passage dealing with the parallax of the Moon, that is, the change in the Moon's position because it is relatively close to the Earth. And it would not be the first time in history that students had swallowed a lecture hook, line, and sinker without scrutinizing their teacher's claims. Despite his starting errors, however, Offusius did manage to calculate some of the later numbers slightly more accurately than Copernicus had done.

The astronomer who eventually discovered that Offusius' criticism was wrong turned out to be none other than Paul Wittich. Tracing the tenuous connection from Offusius to Wittich is a curious but informative story in itself. In the back of the Beinecke Library copy of *De revolutionibus* at Yale, a note describes Wittich passing on to Hagecius a terse list of three errors in Copernicus' text. These three errors exactly match three errors noted in some detail in the margins of Offusius' own *De revolutionibus*. Wittich must therefore have seen that copy (or possibly the one now in Debrecen, the only other one to note all three supposed errors), but he himself did not transcribe the long critique of folio 127 into any of his own four copies of the book. Nevertheless, on folio 127 in his copy (now in the Vatican Library) he made a careful analysis of Offusius' starting claim and showed it wanting, and he also cited the other two rather trivial errors in its margins, adding *secundum Jofrantium* (according to Jofrancus)."*

Finding this convoluted connection to Wittich was totally unexpected fallout from the Offusius case, a stunning confirmation of the "invisible college," that network of sixteenth-century astronomical communications outside formal university instruction. The invisible college comprised tutorial and mentor relationships that transcended institutional

---

*The note in the Yale copy connecting Hagecius and Wittich is in the hand of Johannes Praetorius, who in 1576 had become professor of mathematics at Altdorf, a town just east of Nuremberg. In some way he had gotten the information from Thaddeus Hagecius, a man with broad astronomical interests and connections in the courtly circles of Prague. No *De revolutionibus* with a Hagecius provenance has turned up, but it is tempting to think that at some point he might have owned the Edinburgh copy with the Offusius notes and that he showed it to Wittich.

boundaries. The evidence for such connections is of course now made visible by scholarly sleuthing, but those sociological structures were neither formally codified nor rooted in the universities.

While the Offusius linkages present some of the best evidence for this communication network, it is by no means unique. Another example is a *De revolutionibus* I found in the Ossolineum in Wrocław, Poland. This second edition contains annotations from two different traditions, from Michael Maestlin in Tübingen and from Paul Wittich somewhere in Central Europe (or from the copy of Wittich's notes made by the Scot John Craig). Like all too many of the copies, it doesn't record who made the annotations or where the copying took place; nevertheless, the notes show that, despite the challenges of travel and communications, somehow the messages made the rounds.

WHILE THE identification of Jofrancus Offusius as the Master of the 1550 Radices cleared up a major mystery relating to the copies of *De revolutionibus* that I had examined, there remained a serious puzzlement. Copernicus' only disciple, Georg Joachim Rheticus, who had persuaded the aging Polish cosmologist to authorize the publication of his book, was surely a member of the invisible college, inspiring students and communicating with his contemporaries, yet where was any trace of a family of annotations stemming from his own copy of *De revolutionibus*?

Although Rheticus had been appointed both professor and dean when he returned to Wittenberg after his Polish sojourn, he taught there only briefly before he headed south to Nuremberg with the manuscript of *De revolutionibus*. Perhaps he was still not entirely comfortable at Wittenberg. But his influential friend Philipp Melanchthon busily pulled strings on his behalf. Joachim Camerarius, a distinguished humanist and Greek scholar, was in the process of transferring from Tübingen to become the leading professor at Leipzig. Rheticus must have impressed Camerarius several years earlier when he visited Tübingen before setting out to see Copernicus in Frauenburg. By then a particularly well-informed astronomer, Rheticus soon received an offer to become the number two professor in Leipzig, and at a particularly tempting salary, 140 florins per annum.

(The standard rate was 100 florins.) So Rheticus took the manuscript to Nuremberg in the spring of 1542, got the printing started, and then in the fall left for Leipzig. Camerarius had made something of a cottage industry of writing Greek prefatory poems, and Rheticus asked him to compose an appropriate Greek poem to introduce Copernicus' book.

In April 1543 Rheticus received his first complete copies of *De revolutionibus*. He must have been shocked. Not only was Camerarius' poem omitted, but there was no acknowledgment of his own role in publishing the book. Instead, the book opened with Osiander's anonymous apologia, the infamous *Ad lectorem*.

Rheticus was incensed. With a red crayon he struck out the unexpected *Ad lectorem*. Then he asked Camerarius to pen onto the flyleaf the Greek poem he had composed for the book. The iambic pentameters carried a Platonic dialogue between a philosopher and a stranger, who was ultimately advised not to criticize as the ignorant might do, but to study and then try to do better. Finally, he signed the book as a gift to Andreas Aurifaber, then the dean at Wittenberg. Taking another copy, Rheticus again applied the red-crayon deletion. Without the poem, he inscribed it for George Donner, a canon at the cathedral in Frauenburg (plate 7d). Another similar copy he dedicated to Tiedemann Giese, Copernicus' best friend and the bishop of Chełmno; in fact, he sent two copies to Giese. Camerarius penned his Greek poem into another copy. Was it his own copy, or was it one that Rheticus decided to keep for himself? Three of these copies survive today, and another is attested in the historical record, though it has vanished. Yet none of them could qualify as Rheticus' working copy, one to compare with the magnificent, thoroughly marked-up copy from Reinhold preserved at the Royal Observatory, Edinburgh.

Copernicus himself did not see a completed book for another month after Rheticus received his copies, on account of the extra transit time required to send it to his far-off outpost in northern Poland. The final pages to be sent were of course the unnumbered opening pages, the so-called front matter. Copernicus had by then suffered a stroke and was partially paralyzed. Perhaps he, too, was distressed by the anonymous introduction that Osiander had added. Or, more likely, he might simply have been holding out against the dark angel of death until he could at last see his work completed, and thus satisfied, he died.

Tiedemann Giese described the end in a heartfelt letter to Rheticus. He apologized for the lack of recognition for Rheticus, saying that in the final days Copernicus was forgetful, and he urged Rheticus to demand that the front matter be reprinted without the Osiander piece. Nothing came of his suggestion.

Rheticus carried on as both the astronomy and the mathematics professor at Leipzig. His reputation was such that some of his students from Wittenberg transferred to Leipzig. But once again his wanderlust caught hold of him, and he yearned to visit Italy—perhaps remembering what Copernicus had told him of his own graduate student days there. So in the summer semester of 1545 Rheticus took a leave and journeyed south. And once again what presumably started as a brief trip took on a life of its own, and for nearly three years he was absent.

Among the people he visited was the Italian mathematician Girolamo Cardano. In 1545 Cardano had published with Petreius in Nuremberg the greatest mathematical text of the sixteenth century, his major work, *Ars magna*, wherein he had praised Andreas Osiander for his skill as a proofreader. "He not only knows Latin and Greek," Cardano wrote, "but even mathematics." No doubt Rheticus and Cardano discussed Copernicus during their visit, but could that conversation have led to the aphorism that Cardano included in his *Aphorismata astronomica* that he published with Petreius in 1547? There, under aphorism 69, he wrote, "Indeed, the opinions of Copernicus are not yet well understood, for he barely seems to say what he wants."

Returning from Milan, Rheticus visited Lindau, where he lay seriously ill for five months, then Constance, Zurich, and Basel. Finally, after three years, he was back lecturing again in Leipzig. Did he teach any of his students about the new cosmology? Presumably he continued to be familiar with the technical details, because in 1550 he published an ephemeris for 1551 based on *De revolutionibus*. In the previous year he edited an edition of the first six books of Euclid's geometry. This time was the high-water mark for Rheticus' professorial career.

Then, in the spring of 1551, Rheticus' world came crashing down. He had fallen deeply in debt, to the tune of twice his annual salary, and soon in scandal. There were dark rumors of a drunken homosexual episode involving a student half his age. The irate father of the young man involved brought a lawsuit. In disgrace, Rheticus fled from Leipzig.

Rheticus moved eastward, first to Prague, where he took up the study of medicine (just as his mentor, Copernicus, had done), then to the University of Vienna, and finally to yet another university city, Cracow, which became his residence for nearly twenty years until the last year of his life, when he moved to Hungary. He took with him his work on a massive trigonometric table, sines and cosines to ten decimal places, an accuracy unsurpassed until the modern computer age. The project lagged as his interest in medicine grew, a career shift of a number of upwardly mobile professors. Ever the rebel, he took up the new medicine advocated by Theophrastus von Paracelsus (using chemical drugs rather than the traditional herbs), the radical medical equivalent of heliocentrism.

Only toward the end of his life, especially with the arrival of a young collaborator, Valentin Otto, in 1573, did his enthusiasm for the calculations revive. Rheticus was moved by the obvious parallel: Just as Rheticus, as an eager young researcher, had worked with and encouraged the elderly Copernicus to publish *De revolutionibus*, so, too, did the young Otto play midwife to Rheticus' *Opus Palatinum*, the ten-place table of sines, cosines, and secants.

Clearly, the destiny of Rheticus' personal copy of *De revolutionibus* was one of the mysteries of the entire Copernicus chase. Even if his own working copy didn't survive, surely some students would have made copies. Though Rheticus served as dean and professor in Wittenberg only for one semester, his students included seven men who later became mathematics professors either at Wittenberg or elsewhere in the Lutheran university system. Copies of *De revolutionibus* from four of them survive, and for a fifth we have evidence of the annotations in his book. None of these show evidence of a standard pattern of annotations that could have been copied from Rheticus. Making an ephemeris from Copernicus' work, as Rheticus did in 1550, was not a particularly easy task, because some of the essential numbers were well buried in the text. Highlighting these numbers seems like an obvious teaching aid, and while the copies from Jofrancus Offusius' students show precisely this kind of marginal additions, for example, no copy attests to such activity on Rheticus' part.

This is where the matter stood until 1998, almost three decades into my census research. Then, unexpectedly, a rather interesting second-edition *De revolutionibus* appeared in London. Paul Quarrie, who was aware of my investigations ever since I had turned up at Eton to inspect its two first editions while he was librarian there, had in the meantime become Sotheby's rare book expert. He alerted me to the volume, which had been consigned to Sotheby's, and when I inspected it, I was able to point to a special feature of that copy. When Petreius finished printing the book proper, he immediately printed an errata leaf with the errors on the first 280 pages of the 400-page book. There is a logical explanation why Petreius printed only the first part of the errata. I am guessing that when the last pages went through the press, the corrections for the final part hadn't yet been returned from Copernicus, so on the errata leaf he printed what he had and that was that. Later the remaining corrections came to hand but were never printed, although a few insiders in Wittenberg got access to them. In the course of compiling the census I had found seven copies in which the errors were hand-corrected all the way to the end of the book, not just on the first 280 pages given on the printed errata leaf. The *De revolutionibus* at Sotheby's was an eighth copy with the errata marked all the way to the end.

Paul Quarrie carefully described the book in the auction catalog. He mentioned a characteristic comment on folio 96 that I had recorded in several of the other copies. This was a rather interesting marginalium alongside the place where Copernicus raised the question whether it was the Sun that was the center of the universe or the center of the Earth's orbit, the place in the text that Kepler and Maestlin must have specifically discussed.

I coveted the copy coming up in the auction, but I didn't have the cash to back up a winning bid. It went to a French collector who wasn't particularly interested in annotations, so by and by I offered him my own copy of the second edition plus a cash premium in order to make an ex-

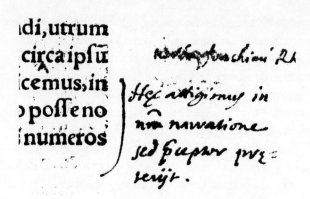

*An annotation copied from Rheticus' original note, possibly
by his assistant Valentin Otto, author's collection.*

change. We rendezvoused at a modest French restaurant near Place Den-
fert Rochereau in Paris, had a pleasant conversation about book collecting, and consummated the swap. I carried my acquisition over to the
nearby Observatoire de Paris and showed it to Alain Segonds, a colleague who had been unusually helpful with the census. Then I took a
closer look at the annotation on folio 96, and I was stunned.

In my previous inspection of the book and in glancing at the transcription in the Sotheby's auction catalog, I had simply assumed that the
annotation matched several in the other, related copies. What I hadn't
noticed was that the annotation was in the first person. It read, "We
touched on this in our *Narratio* but my teacher skipped over it," and the
note was specifically attributed to Georg Joachim Rheticus. It was not in
Rheticus' hand, but it was surely a verbatim copy from Rheticus' own *De
revolutionibus*. I hadn't found Rheticus' own book, but here was a transcription of his sparse notes and his list of corrections, including a few
not found in the other copies with the extended errata list. The second
edition had been published in 1566, and several years later Valentin
Otto had gone to Hungary to work with Rheticus, so quite possibly
what I now have is a copy Otto made at that time. Alas, Rheticus' annotations, at least those transcribed into my copy, scarcely offer intrigu-

ing insights, at least not the way Maestlin's or Kepler's marginalia do. Rheticus clearly understood the astronomy at a deep level, but it seems he lost interest in following up on the technical details, and the copy of his annotations shows that only too vividly.

Did Rheticus dust off his interest in heliocentric astronomy when his young visitor arrived? Who can guess what stories the sixty-year-old Joachim told the thirty-year-old Valentin about his own apprenticeship a quarter century earlier with the master cosmologist? In those bygone days Copernicus and Rheticus must have discussed astronomy, medicine, student days in Italy—and possibly even astrology.

Chapter 12

PLANETARY INFLUENCES

DURING THE great Copernican Quinquecentennial of 1973, two distinguished scholars had been assigned a private limousine to get them from Warsaw to Copernicus' birthplace, Toruń. Edward Rosen, the dean of Copernican studies, and Willy Hartner, Europe's leading historian of the exact sciences, emerged from the car no longer on speaking terms. Hartner had had the audacity to suggest that Copernicus and Rheticus could have discussed astrology. After all, in the *Narratio prima* Rheticus had entitled one section "The kingdoms of the world change with the motion of the eccentric" and added that "this small circle is in truth the Wheel of Fortune." Surely, he would not have included the statement without having talked about it with his mentor. To Rosen the very idea of such a conversation was anathema. To him Copernicus was the model modern scientist, unpolluted by such notions as planetary influences.

By today's historiography, of course, Rosen's view was hopelessly anachronistic. Copernicus lived in an era when astrological ideas permeated academia. The astronomy curriculum was designed to teach advanced students the use of planetary tables so that they could calculate the positions of planets needed for constructing a horoscope that would show the aspects of the sky at a patron's birth. Cracow University had two astronomy professors, one in the arts faculty and the other in the medical faculty, the latter expressly for teaching the doctors-to-be how to use the stars for medical prognostications. Domenico Maria Novara, the astronomer with whom Copernicus boarded while he studied law in Bologna, published annual astrological prognostications, something Copernicus could scarcely have ignored. And when he returned to Italy to study medicine at Padua, he surely was exposed to more astrological thinking.

A century and a half earlier Geoffrey Chaucer had spiced his *Canterbury Tales* and his *Troilus and Crysede* with planetary configurations that held keys to the twists of fate in his stories, and if anything, the astrological ethos had only intensified since his time. Even today our language contains fossil remnants of a sidereal and planetary vocabulary: *consider, ascendancy, disaster, jovial, martial, venereal, mercurial, saturnine,* not to mention the names of the days of the week.*

Copernicus was born 19 February 1473 at 4:48 P.M. This fact would probably be unknown to us except that it is preserved in an early manuscript horoscope found in the Bayerische Staatsbibliothek in Munich. It seems unlikely that Copernicus' mother had a clock beside the birthing bed or that the time would have been recorded so accurately. In fact, Renaissance astrology guides point out that the first step in constructing a horoscope was usually to retrodict the time of the client's birth. An elaborate process of deducing the moment of conception by examining the phase of the Moon nine months earlier, and then working forward in time, theoretically allowed the astrologer to construct the missing information. Since a critical feature of each horoscope is the degree of the zodiac that was just rising at the moment of birth (the so-called ascendant), and because this changes on average every four minutes, a fairly precise time of birth is required. The chances are extremely high that Copernicus' birth moment was simply calculated.†

All the available biographical information on Rheticus reveals his passion for astrology. Curiously, there is not a shred of evidence that

---

Consider = cum sidera = "with the stars"; *disaster* = "against the stars" (ill-starred); et cetera.

†In Johannes Kepler's manuscript legacy there are, among many others, two horoscopes he drew up for himself, one for his birth and the other for his conception. Once when I projected slides of both side by side in my class, a young woman in the second row raised her hand to inquire why there were only seven and a half months between the two horoscopes. "Oh," I replied, "Kepler chose his parent's wedding night for the date of the conception." Needless to say, this remark completely cracked up the class.

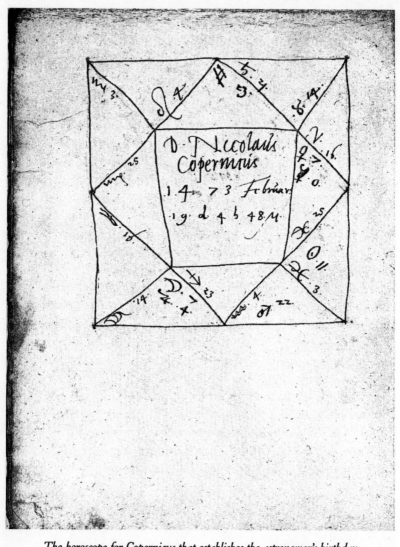

The horoscope for Copernicus that establishes the astronomer's birthday.
Bayerische Staatsbibliothek, Munich, Cod. Lat. #27003, folio 33 verso.

Copernicus had any interest in the subject, even though he could hardly have avoided learning the standard rules of its practice. Given the ethos of the times, Rheticus and Copernicus must certainly have discussed the topic. Copernicus was surely not naive; he must have realized that astrologers would constitute a good fraction of the market for his treatise.

WELL IN TUNE with the realities of the sixteenth century was René Taylor, director of the art museum in Ponce, Puerto Rico, who had come to me with an unusual request several years before my Copernicus chase began. Could I draw up a horoscope for the laying of the cornerstone of the Escorial Palace north of Madrid, an event that took place on the Feast of St. George in April 1563? I was naturally curious as to what lay behind his request. It turned out that he had a list of books in the library of Juan de Herrera, architect of the Escorial. There were few books on architecture but many on magic, astrology, and astronomy. Taylor figured that Herrera would naturally have used astrology to choose the time of day, and perhaps even the day itself, for the cornerstone ceremony. Because I was then in my spare time computing planetary positions from ancient Babylonian times to the present, I figured I could easily help him test his hypothesis.

Little did I then appreciate that every major Renaissance astrologer seemed to have his own way of dividing up the sky into the so-called astrological houses. Furthermore, for historical studies it was necessary to know not where the planets really were but where the astrologers thought they were (which was often quite another thing). In agreeing to help Taylor, I got myself in deeper than I had intended, but eventually I came up with a credible sixteenth-century horoscope, and I helped Taylor find a real astrologer, someone with a genuine thirteenth-century mind,* to interpret it and to convince him that the time and day were deliberately selected to be astrologically propitious.

---

*The expert was an authority on Islamic art of the Middle Ages, who was well in tune with that era.

Herrera's library had included not one but two examples of Copernicus' *De revolutionibus*. After my quest began, I looked forward to the opportunity to inspect them in the Escorial itself, and in 1977 I finally had the occasion to visit the splendid library with the astronomical frescoes that had been at the heart of Taylor's researches. In the course of chasing after copies of Copernicus' book, Miriam and I had visited some spectacularly decorated rooms. The library at the Melk Abbey in Austria comes to mind, as do both the Strakhov Monastery and the Clementinum in Prague. The original reading room of the Vatican Library is pretty grand as well, and the Österreichische Nationalbibliothek in Vienna is close to being in the same league. Alas, we didn't actually sit in any of these wonderful spaces, including the Escorial, to study *De revolutionibus*. Those ornate halls now function as art galleries, and the reading rooms are not so stately. So for total experience, the Wren Library at Trinity College, Cambridge, and Duke Humphrey's Library at the Bodleian in Oxford take the laurels.

Still, we took a special delight in actually seeing the Escorial images. The vaulted 175-foot-long arcade provided space for strategically positioned showcases of rare books and manuscripts as well as an ornate armillary sphere, and the hall was lined with Herrera's specially designed bookcases of ebony, cedar, orangewood, and walnut. The frescoes individually were not particularly memorable art—the conservative King Philip II, patron of the palace, placed the flamboyant paintings by El Greco elsewhere—but the total effect in the hall was stunning. The arched panels depict the seven liberal arts, beginning with Grammar and Rhetoric at one end, Logic, Arithmetic, and Music in the central three vaults, and Geometry and Astronomy curving over the armillary sphere. There in the bay for Astronomia is Alfonso the Great, patron of the astronomical tables that Reinhold's Copernican-based *Tabulae Prutenicae* ultimately replaced in the mid–sixteenth century. Facing Alfonso on the other side of the arcade is Euclid, and like the Castillian monarch, he holds astronomical symbols that, according to Taylor, were connected to the horoscope of Philip II. And under the bay is another

The main gallery of the Escorial Library, with its frescoes of the seven liberal arts.

astronomical fresco, with Dionysius the Areopagite observing the eclipse at Christ's passion.*

The library itself holds one of the world's largest collections of medieval manuscripts, being surpassed only by the Vatican's holdings. In the austere reading room I took a look at the library's copies of *De revolutionibus*. Its first edition is in a green pigskin binding blind-stamped with the arms of Philip II—that is, an impressed but uncolored design. Apparently the king had bought the copy early on, in 1545 when he was eighteen. He didn't write in it, and there is no way to know if he actually read it. The king was a formidable collector of books, and this one became part of a gift

---

*There couldn't have been an eclipse at that time. Jesus was crucified the day after Passover, which is locked to a lunar calendar so that it falls near the time of full moon. Solar eclipses occur only at the time of new moon.

of more than 4,500 volumes he presented to the Escorial monastery library in 1576. I was disappointed not to find direct evidence there of either of the copies Juan de Herrera was known to have owned. The Escorial suffered a horrendous fire in 1671, and if Herrera's volumes were first editions, they might have perished then. There are, however, two unattributed copies of the second edition in the Escorial collection. If the two copies Herrera owned were of the 1566 Basel edition (entirely possible, for he was also collecting books after that time), one or both of these copies now in the Escorial might have originally belonged to the architect and escaped the fire. We will probably never know whether the Escorial architect actually read the Copernican tome, because these copies have no annotations, and I never found any other copy associated with him.

As I EXAMINED many annotations in libraries throughout Europe and North America, virtually always in Latin, I gradually learned to distinguish between various national writing styles. But no matter the nationality, there are both neat, lucidly legible hands and some so messy or idiosyncratic that reading them provides a test of wit and patience. One scrawl on a title page in the National Library in Rome turned out to be relevant to astrology, but for many days it defied both me and my colleague Jerzy Dobrzycki as we tried to decipher it. Early one morning, when I was still half asleep in bed, the reading came to me. Many seemingly brilliant insights occur when I'm in such a semiconscious state, only to dissolve into triviality in the clear light of dawn, but this rare exception rang true. Dobrzycki was visiting me at the time, and he expressed considerable skepticism that I had managed to crack the inscription, but he was soon obliged to concede. I was quite pleased with my success, since I could so seldom best him at a Latin transcription. The phrase read, "Vidit P. R^d Inquisitor inde Corrigatur si qua erant astronomiae judiciare die 2 apl 1597" (The Reverend Father Inquisitor saw it so that it would be corrected if there were any judicial astrology, on 2 April 1597). In other words, the book passed muster because it contained no astrology. Ed Rosen would have been very pleased.

Sometimes an annotation in a cryptic hand seemed almost impossible to decipher. For example, Dobrzycki and I puzzled for a long time over the following note in a copy of *De revolutionibus* at the National Library of Russia in St. Petersburg (here considerably enlarged).

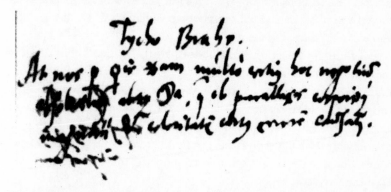

We recognized the third symbol in the first long line under the name Tycho Brahe as the letter *p* with a horizontal bar through the tail, the standard Latin abbreviation for *per*, but what came next baffled us. I finally showed it to Emmanuel Poulle, the distinguished French paleographer who had helped in many ways with the census, and he almost instantly read it as "At nos per Veneris stellam multo certius hoc negotium observavimus alioquin Luna,* etc." With that I felt rather like a kindergartner, for Jerzy and I had often encountered the planetary symbols in the notes, but in this context we had totally missed the traditional symbol for Venus. These symbols are tiny planetary logos, depicting a recognizable characteristic of each member of the Greco-Roman planetary pantheon:

$$\text{☿} \qquad \text{♀} \qquad \text{♂} \qquad \text{♃} \qquad \text{♄}$$

---

*(But we have observed this business with more certainty via the planet Venus than using the Moon, etc.) The note refers to Tycho's method of comparing the position of the Sun with the stars, which is difficult because they are not visible at the same time. Tycho connected them by measuring the distance from the Sun to Venus (which is visible in the daytime), and at night connecting Venus with the stars. In the text next to the annotation Copernicus described using the Moon for the same purpose. Tycho's alternative method gave improved accuracy.

The symbols for the five naked-eye planets are given here in their traditional order, beginning with Mercury with its truncated caduceus, the serpent-entwined staff of the swift messenger of the gods. Venus' mirror is obvious (the second symbol), as is the spear and shield of Mars (the third). Jupiter's symbol is a stylized thunderbolt, while Saturn's is the scythe of Father Time (the last in the row).

In hindsight, the printing methods of the 1970s that I considered for the *Census* in its early stages seem very primitive by today's standards. Many entries were typed and retyped by my secretary as additional details and identifications became available. Although today I can hardly remember what a Spinwriter is, for some years it was the device of choice for a reasonably legible computer printout. In retrospect, some of the world's ugliest books since the invention of printing with movable type were produced in that decade.

As larger and faster computers arrived on the scene, computer-typesetting became an everyday reality. Then I realized that for the typography of the *Census* I would need the symbols both for the planets and for the signs of the zodiac—zodiacal signs because astronomers in those days frequently designated astronomical longitudes by a system in which the ecliptic* was divided into twelve equal segments of thirty degrees, for example ♈ 14°. Here the symbol for Aries derives from the horns of the Ram. I wanted my *Census* to have symbols that reflected their historical roots, so I commissioned the California typographer Kris Holmes to add these characters as well as the planetary symbols to her Lucida computer font. I sent her images from sixteenth-century astronomical tables, and here is the typography we agreed on.

---

*The ecliptic is the great circle of the Sun's path through the zodiac. The paths of the Moon and planets are slightly tilted with respect to the ecliptic, each orbit crossing the ecliptic at two points called the nodes, separated by 180°. An eclipse can take place only when the Moon is near its node, crossing the ecliptic (with the Sun also at a lunar node), hence the name *ecliptic.*

♈ Aries	♋ Cancer	♎ Libra	♑ Capricornus
♉ Taurus	♌ Leo	♏ Scorpio	♒ Aquarius
♊ Gemini	♍ Virgo	♐ Sagittarius	♓ Pisces

The meanings of some of these symbols are pretty obvious: the rippling waves for watery Aquarius, the balance for Libra, the arrow for Sagittarius the Archer. Others are more abstract, such as the bushy mane of Leo the Lion, the paired lines for Gemini the Twins, or the horns and face of Taurus the Bull. (Our alphabet may have derived from a denser set of astronomical symbols beginning with Taurus; tipped ninety degrees, the Taurus symbol becomes an alpha.) The two most confusing are the symbols for Scorpio, with the pointed stinger in the scorpion's tail, and Virgo the Virgin, which carries the standard medieval abbreviation for Maria, a capital *M* with the crossed tail, a shorthand indicating additional letters.

As with the abbreviation for Maria, the early annotators frequently designated omitted letters with a short line above, below, or behind the nearest letter or over the entire word. The commonest abbreviation is a bar over a vowel, indicating a missing *n* or *m*. The system works up from there. For example, *oā* stands for *omnia*, or *rō* can represent *ratione*, something of an all-purpose word in the technical Latin context, meaning everything from "reason" to "theory" to "thought." In working with the Latin script, I quickly learned to distinguish between *p* for *per* and *ρ* for *pro*. It was much harder to keep track of *q* for *qui*, *q̃* for *quae*, and *ɋ* for *quod*. In the *Census* these words are always spelled out, so fortunately I didn't need a virtual type box quite as large as the one Petreius used for setting *De revolutionibus*.*

The use of abbreviations was generally the compositor's option in the

---

*Not counting punctuation, numbers, Greek letters, or the zodiacal signs and planetary symbols, the Nuremberg printer used just over thirty special symbols or ligatures in addition to twenty-three lower-case letters (no *j*, *v*, or *w*, which do not occur in Latin) and twenty-three uppercase letters (no *J*, *U*, or *W*).

sixteenth and seventeenth centuries. When it was necessary to squeeze the words to get them to fit in a line, he would select more abbreviations from his type box. A few lines later, with a more relaxed spacing, the word might well be spelled out in full. It's particularly interesting to see this process at work in Galileo's *Sidereus nuncius*. The final page is thoroughly sprinkled with abbreviations as the typesetter exercised full control to make sure that the text didn't run over by a few lines onto another sheet. That happened in *De revolutionibus*, too, at the end of Book V where the typesetter jammed in far more than the average number of abbreviations. The effect also happened in the middle of the signature* marked *s* in or-der to finish a chapter before four pages of tables. Probably a virtuoso compositor took over, squeezing in abbreviations like *tpe* for *tempore* or *qñ* for *quando*.

IN 1974 I made two complicated, adventuresome book-hunting trips to Europe, adding a few dozen more Copernicus copies to the census. At Hertford College in Oxford I encountered a particularly eccentric librar-ian who suspected me of being a fraud, but he did show me Hertford's unannotated second edition, which had been owned by Thomas Finck, the seventeenth-century physician and mathematician who added the words *tangent* and *secant* to the trigonometric vocabulary. That April trip eventually carried me on to Egypt, where the astronomers had been given funding for a Copernican quinquecentennial commemoration, so they en-listed me to add an international note to their one-year-late affair. But in be-tween these stops I went to a conference in Capri that led to a quite unexpected finding about Galileo.

I knew that Capri was a famous resort destination, but I was quite sur-prised to discover that the island had no beaches. Just getting there proved to be both a figurative and literal cliff-hanger. The plane from London to Rome was an hour late, and by the time I could catch a high-speed train

---

*A signature in *De revolutionibus* comprised four printed leaves, or eight pages. Each signature was coded with a sequential letter to assist in the assembly of the book.

to Naples, it was one scheduled to arrive five minutes after what I had been led to believe was the last ferry to the island. I had nervous visions of hiring a fishing boat at some exorbitant rate to take me there, little realizing what a substantial distance was involved. Fortunately, the information was wrong; by great good luck I had got out at the right train station, the one next to the ferry terminal, and I just had time to catch what really was the last ferry. I reached the storybook island at dusk, ascended the towering cliffs by funicular, and was soon in the company of some of the leading historians of the scientific Renaissance.

My role at the conference was to comment on a paper presented by Guglielmo Righini, director of the observatory in Florence. Righini examined in detail Galileo's early drawings of the Moon in an attempt to date when they were made. In 1609 Galileo had learned about a Dutch spyglass that was being sold in the major cities of Europe; he figured out how it could be done and effectively turned what had been a toy into a scientific instrument, with which he discovered the craters on the Moon. His book announcing his discoveries, *Sidereus nuncius*, or "The Sidereal Messenger," was published in March 1610, so the possible dates of his observations were fairly limited. Strangely enough, no one had attempted to pin specific dates onto the two surviving sheets of Galileo's observations, probably because no one before Righini had believed that they were accurate enough to warrant such an investigation. It fell to me to point out that there were serious problems with the images published in *Sidereus nuncius*, which varied in critical ways from the original ink-wash drawings, and that the dates Righini picked were not necessarily unique.

Although Righini's analysis was flawed, his paper provided the catalyst that ultimately established a precise dating of Galileo's lunar observations and thereby gave an accurate chronology for the swift genesis of his astronomical discoveries. Galileo undoubtedly found the craters on the Moon before he was prepared to record them, but having decided that his discovery was worthy of publication, he subsequently equipped himself with ink, brushes, and a sheet of special artist's paper, and on the evening of 30 November 1609 made two careful depictions of the cratered lunar surface.

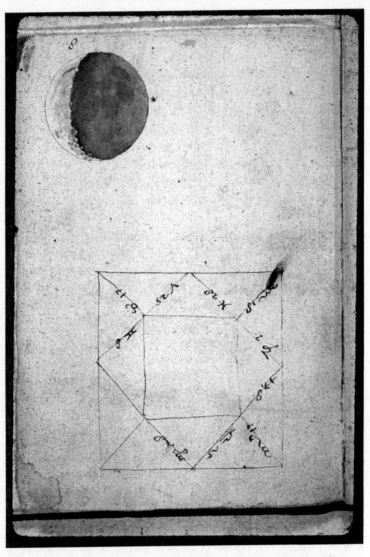

Galileo's ink-wash drawing of the Moon on 19 January 1610 and his uncompleted horoscope for Cosimo de' Medici; a completed horoscope is on the other side of the sheet.

At four further times throughout December he added images to his sheet, making six drawings in all. But in the week of 7 January 1610 another series of observations, not of the Moon but of Jupiter, suddenly gave an urgency to his publication schedule. After discovering four companion stars revolving around Jupiter, and filled with excitement but fearful of being scooped, he rushed to publish his observations. His *Sidereus nuncius* was in print in just over six weeks after he had delivered the first installment of his manuscript to the printer in Venice—an extraordinarily quick turnaround even by today's standards.

In a letter to Kepler written in 1597, Galileo had allowed that, privately, he accepted the Copernican cosmology. In public, however, for more than a decade Galileo apparently never whispered a hint of his radical beliefs. That all changed in 1610 with the publication of *Sidereus nuncius*. Some critics had resisted the idea of a moving Earth, asking how an Earth in orbit around the Sun could keep the Moon in tow. To them, Galileo pointed out that Jupiter, which everyone agreed was in motion, managed to retain its companion moons as it moved across the sky—a powerful Copernican counterargument to the objectors. From this point on, Galileo became ever more open in his defense of the Copernican system, apparently stimulated by the astonishing novelties his telescope had revealed.

But Galileo had a second agenda in writing *Sidereus nuncius*: He was keen to give up his professorial post in Padua to become the mathematician and philosopher to Grand Duke Cosimo de' Medici in Florence, and to that end he dedicated his book to Cosimo and he named the Jovian moons the "Medicean stars." In this plan he succeeded, and Florence became his home for the rest of his life.

I made a second trip to Italy in July 1974, and in Florence I discovered what turned out to be Galileo's secret weapon in getting the job at the Medicean court. At the time of the Capri conference I had not actually seen Galileo's original ink-wash drawings of the Moon. Two pages of the lunar images had been carefully reproduced in the so-called National Edition of Galileo's works and correspondence, published in twenty volumes at the turn of the twentieth century, and it was very convenient to use those reproductions in preparing my commentary on Righini's paper. But with my curiosity sparked, I took advantage of a visit to Florence to examine not only Galileo's sparsely annotated copy of *De revolutionibus* in the National Library but also his astronomical manuscripts. I had long supposed that Galileo was not the sort of astronomer who would have read Copernicus' book to the very end. Even on that seminal evening with Jerry Ravetz, when we had speculated how few early readers of *De revolutionibus* there might have been, we had been reluctant to include Galileo in the list of readers. Unlike Reinhold or Maestlin or Kepler, he was not interested in the details of celestial mechanics. Still, when I saw the copy in Florence, my reaction was one of skepticism that it was actually Galileo's copy, since there were so few annotations in it apart from the standard censorship decreed by the Inquisition in 1620. Eventually, as I became more familiar with Galileo's hand, I realized that my skepticism was unfounded and that it really was Galileo's copy.

The Galileo manuscripts, on the other hand, proved to be quite fascinating. To my surprise, I discovered that the reproductions of the sheets of lunar drawings in the National Edition were not entirely complete. A single drawing of the Moon on a second sheet, now dated to 19 January 1610, had been published, but an astrological horoscope that shared the page was nicely suppressed. Clearly, it would have diminished Galileo's heroic status as the first truly modern scientist to admit so conspicuously that he was capable of drawing up horoscopes.

I asked Righini's wife, Maria Luisa Bonelli-Righini, director of the History of Science Museum in Florence, for help in obtaining color slides of the horoscopes (for there was another on the other side of the sheet). They arrived just in time for me to add a postscript to my Capri paper, and there I pointed out that one of the horoscopes could be dated to 2 May 1590 from the positions of the planets it contained. Afterward I felt rather dim-witted that I hadn't taken the next step. Righini, alerted to the horoscopes by my slide request, promptly realized that the date was Cosimo's birthday, and that Galileo had drawn up a birth horoscope for his prospective patron. Indeed, in the dedication to *Sidereus nuncius,* Galileo dwelled on the theme of Jupiter's position in the horoscope, writing with obsequious flattery, "It was Jupiter, I say, who at your Highness's birth, having already passed through the murky vapors of the horizon and occupying the midheaven and illuminating the eastern angle from his royal house,* looked down upon your most fortunate birth from that sublime throne and poured out all his splendor and grandeur into the most pure air, so that with its first breath your tender little body and your soul, already decorated by God with noble ornaments, could drink in this universal power and authority." Astrology never again played a public role in Galileo's work, unlike Kepler's, which included a pamphlet entitled *The Sure Fundamentals of Astrology* and another defense where on the title page he urged critics not to throw out the baby with the bathwater, saying that he was searching for a few kernels among the dung of traditional astrology. Nevertheless, astrology was part of the ethos of the times, and it is surprising that there is nary a hint of it in *De revolutionibus.* Nor is there any trace of interest in astrology in anything else that remains from Copernicus.

---

*The two most important zones, or "houses," in a horoscope are the ascendant (the part of the ecliptic circle about to rise over the eastern horizon), here called the eastern angle, and the midheaven (the part of the ecliptic circle about to cross the meridian). In Cosimo's horoscope, Jupiter was in the midheaven, while the sign of Sagittarius, the so-called day domicile of Jupiter, was in the ascendant.

Chapter 13

SOPHISTICATED LADIES

LONG BEFORE I met Alexander Pogo I heard stories of his extraordinary career. Born in St. Petersburg, Russia, in 1893, he had begun engineering studies in Liège in 1911. When the Germans invaded Belgium during World War I, he became a prisoner. After the war he finished his degree, but rather than returning to his native land, by then the Soviet Union, he went to Athens, where he landed a job measuring the fallen column drums of the Parthenon on the Acropolis, preparatory to reerecting some of them. Later he emigrated to the United States, where he earned a doctorate in astronomy from the University of Chicago. In the meantime, he had become fluent in eight languages. With this unusual background and his facility with languages, he became an assistant to George Sarton, the man considered the father of modern history of science.

Sarton's office was in Harvard's Widener Library, though his salary, as well as Pogo's, was paid by the Carnegie Institution of Washington rather than by Harvard. Those who knew Pogo at Harvard remembered that he had his own pet peeves, such as the fact that the great *Canon of Eclipses* compiled by Theodor von Oppolzer in the 1880s omitted the so-called penumbral lunar eclipses, when the Moon was touched only by the outer shadow of the Earth.

As Sarton approached his retirement, the Carnegie Institution, which felt some responsibility for Pogo, decided that he could play a useful role as librarian for the Mount Wilson and Palomar Observatories, which it funded in California. Thus, in 1950 Pogo transferred from Cambridge to Pasadena, leaving in time to miss the famous "I go Pogo" riot when Harvard students tangled with local police as they got a bit too rambunctious waiting for an appearance by cartoonist Walt Kelly, who was then spoofing the 1952 American presidential elections in his *Pogo* comic strip.

Not only did Alexander Pogo become responsible for the Observatories' library, but he became de facto rare book librarian at the California Institute of Technology. Dean Earnest Watson was building a rare science book collection for that institution, and Pogo took me to see it in 1972 when I came searching for copies of *De revolutionibus*. With a touch of pride he told me that Cal Tech's copy of the first edition was not the initial one received from Dawson's of Pall Mall, an eminent rare book firm in London. When the first copy came, he inspected it carefully, and his eye caught a page that was wrong, a leaf from a second edition inserted among the leaves of the first edition. This substitution was possible because the 1566 second edition was virtually a page-by-page reprint of the first. Each signature of four leaves ended on exactly the same word in both the 1543 and 1566 editions. The easiest giveaway was the fact that the larger initial letters of the chapters differed in the two editions, but there were other differences; for instance, the first edition labeled the diagrams with lowercase letters, whereas the second always used capitals. But the similarity of the pages was such that they could escape notice with a casual inspection. Thus Pogo had every right to be proud of his acute observation. The book went back, and in due time a replacement arrived. But, as it turned out, this was not to be the end of the story.

Important sources for documenting the movement of rare books are the specialist dealer's catalogs, and here I was very lucky to inherit an extensive collection going from the 1930s into the 1960s. These catalogs had been obtained by C. Doris Hellman, a historian of science who concentrated on Tycho Brahe and who was a collector of modest scale but considerable expertise. Included among her several hundred catalogs was an almost complete set from Dr. Ernst Weil, a man I never met but who was probably the first independent book dealer specializing in early science. Collectors like Lord Crawford, who formed important astronomical libraries in the nineteenth century, relied on generalist dealers to acquire their most precious treasures. In particular, Lord Crawford depended heavily on Bernard Quaritch in London, who dealt across the board in rare books. Not until rare science collecting became less rare was there a niche for a specialist dealer, and Weil seized this opportunity.

He first worked with the old firm of Täuber in Munich, but in 1933 he emigrated to England, one of many refugees escaping the Nazi tyranny. There he became a director of the distinguished E. P. Goldschmidt company. His initial catalog, "Classics of Science," offered a copy of Galileo's *Dialogo* (1632) (with the claim that "t h e f i r s t e d i t i o n h a s b e c o m e e x c e e d i n g l y r a r e," rather inflated considering that the *Dialogo* is the least expensive of the great astronomical classics), Kepler's much rarer *Astronomia nova* (1609), and "Copernicus' first published book," the *De lateribus et angulis triangulorum* (1542) (again misstated, because in 1509 Copernicus published a now virtually unobtainable translation of moralistic letters by the Byzantine epistolographer Theophylactus Simocatta).*

A decade later Weil struck out on his own, in 1943 issuing his first catalog under his own name. For ten pounds you could buy an autograph manuscript by Isaac Newton, while four pounds, ten shillings would get you the astronomical tables by the female astronomer Maria Cunitia (1650), a book I most recently saw offered for $13,000. In catalogs 4 and 6 he offered a second-edition *De revolutionibus*, with just enough information for me to identify it with a copy now in a private collection in California. And in catalog 22 there was a first edition of the book containing the cryptic clue "with duplicate stamp of a well-known library." Probably this was the duplicate released from Harvard's Houghton Library. I had discovered that copy in 1976 when I visited the private collection of Henry Posner, a retired advertising executive in Pittsburgh.† His collection is now at the Carnegie-Mellon University.

---

*Not a great contribution to letters, nor a very sound translation, this publication was Copernicus' route to learning Greek.

†Posner told me that many years earlier he had been shown a novelty, a small neon bulb. "That gives bright light," he said. "What are you going to use it for?" He was told it worked just fine for testing automobile spark coils, but his bright idea was to make neon advertising signs. He remarked that Pittsburgh had some of the oldest neon signs in existence because in the early days the anodes were made of heavy platinum. With his resulting fortune he was able to acquire a formidable collection of rare books.

Weil's nineteenth catalog contained a notice that raised my pulse rate: He proposed making a census of Copernicus' *De revolutionibus*! "I hope to be able to publish one day a census for which I have collected data for a long time," he wrote. His census never emerged in print, but he was obviously very interested in the subject because, as the pioneering and leading rare science dealer, he had bought and sold a number of copies of the first edition. Weil was no longer alive when I started my own quest for copies of Copernicus' book, but when I inherited the copy of his catalog 19, I was seized with curiosity. Had he actually started a census? Might it be among his papers somewhere?

It did not take long for me to discover that Dr. H. A. Feisenberger, a bookman who worked for part of his career in Sotheby's book department, had inherited Weil's records; the most important for my purposes was a workbook in which Weil had kept track of some of the more famous science titles as they passed through the market. Weil had made notes, often replete with insider information, on the sale of some thirty first editions of Copernicus' book, not enough for a serious census but nonetheless highly interesting. And though the Cal Tech *De revolutionibus* had come through Dawson's, it was clear that Weil was the real source.

Weil's notes were terse but immensely illuminating. Many of the copies he listed I already knew about because I had worked carefully through the auction records, but certain mysteries remained. In 1950 Christie's had auctioned the "Hurn Court Library" with a first edition that lacked folio 97, but no book answering to that description had turned up in my searches. And in his catalog 19, Weil had offered "the Arthur Ellen Finch copy," but no book matching that provenance had been found.

Weil's workbook was an eye-opener. I found the following entry: "1951 Bought with Scheler Finch copy, last leaf in fac[simile] + Sotheran-Christie copy lacking 1 l[eaf] Made Finch copy fine. Cat. 19, now Calif. Inst. Techn. (Watson tells me, 8/54)." Apparently, Weil, in league with a major Parisian dealer, Lucien Scheler, had bought an incomplete copy at a 1950 Christie's auction, and at about the same time he had purchased privately the so-called Finch copy. What happened thereafter became clear: Weil completed the Finch copy, which lacked a genuine final leaf, using the imperfect Christie's copy, without realizing that the Finch copy had another defect as well, a leaf from a second edition. Dawson's of Pall Mall bought the book from him and shipped it to Cal Tech. When Pogo's sharp-eyed inspection revealed the fault, the book came back. Dawson's returned the book to Weil, who promptly moved the necessary folios to fix up the other copy instead.

Several lines later Weil's notebook read, "9/8/54 made up Sotheran-Christie copy in English XVIIth calf. (now Calif. Inst. 8/55)." Among booksellers and collectors there is a long tradition of "making a copy right." This procedure, like rebinding an old book, can be a bibliographical tragedy when historical information is lost, but it can also sometimes be an intelligent step. In any event, making good books out of bad books is a rather commonplace occurrence.

Even I have been a marriage counselor for the occasional pair of handicapped books. Tucked away in an upstairs rare book nook at Allen's in Philadelphia was a rather horrid copy of Tycho Brahe's 1602 book of instruments. Tastelessly bound in modern library-buckram, it lacked six leaves. In their place were photographs on double-weight photographic paper, vintage late 1920s, altogether a rather repugnant assembly, but I mentally filed away this information. Some months later Bruce Ramer, a New York dealer, lamented to me that he had just purchased, sight unseen at a German auction, what turned out to be another wretched copy of the same book, heavily browned and its title page in shreds. "Measure the page size," I told him, "and maybe I'll take it off your hands."

It turned out that the very best pages in Ramer's dilapidated copy were precisely the ones missing in the copy at Allen's, so I bought them both, had the browned pages washed and slightly bleached at the Fogg Art Museum's paper conservation laboratory, and put together a complete book. While I was at it, I asked an English bookbinder if he happened to have any seventeenth-century boards lying around. Luckily, he had a pair of old calf covers of exactly the right size. My *Astronomiae instauratae mechanica* (Machines for the Reform of Astronomy) will never equal the market price of an original unaltered copy, but it cost only a fraction of an unaltered edition.

I now own what is euphemistically called a scholar's copy and is technically known as a "sophisticated" copy. Given today's primary use of the word *sophisticated* (refined, urbane, cultivated), this may at first glance seem like a contradiction in terms. However, if you consult the *Oxford English Dictionary*, you will find that the original use of the word is "altered from primitive simplicity; not plain, honest, or straightforward." Indeed, wearing makeup covers up the defects, and that's sophisticated.

Armed with the insider knowledge from Weil's workbook, in 1976 I wrote to Alexander Pogo, asking him to inspect folio 97 of the Cal Tech copy. He was at first somewhat indignant, replying that the sophisticated copy had been exchanged for another copy. But when I pressed him, he made a careful inspection and wrote back a rueful analysis of a coffee splash on the replacement leaf.

In other words (I assumed) the replacement page had been subtly colored to match the beige tone of the rest of the book. Actually, I didn't really understand what he meant until a number of years later, when I had an opportunity to return to Cal Tech. By then there was a special reading room for rare books, including a fine collection of Newtoniana given by Pogo. I had neglected to bring along the folio number of the replacement leaf, so it was very difficult to spot the alteration. When I finally found it, I was amazed to see that a stain had probably been placed on several pages to make the added leaf appear to be integral to the original state!

All this left me puzzled as to what had happened to the Finch copy, the one that had originally been sent to California but returned after Pogo's inspection. Could it have gone back to Scheler, who had half ownership in the copy?

Eventually, in 1982, a copy surfaced at the Scheler firm in Paris, by then managed by Bernard Clavreuil. "Oh, it's a copy that's been around here for about thirty years," he told me. Precisely what happened to it is not entirely clear. Apparently Clavreuil bought yet another imperfect copy, and probably by now the leaves have been even more thoroughly shuffled. On a visit to the shop, Clavreuil handed me the sophisticated copy. Naturally, I surveyed the volume very carefully. I remarked that it evidently contained a number of facsimile leaves. "How could you tell?" Clavreuil asked curiously.

I had noticed that the book had been censored according to the Inquisition's instructions, which meant that corrections were specified on eight different pages. Yet in Clavreuil's copy not all the expected places displayed the censor's tracks. The pages lacking the censorship marks were clearly replacements.

"Well," I said, "this is a copy that has been censored according to the standard instructions from Rome. But on the facsimile leaves the censorship is missing." This was true, but mostly I was just letting him know that forgeries give themselves away by a variety of clues. The real detection came from the watermarks.

WHEN A SHEET of paper was made in the sixteenth century, the rag fibers were lifted out of the slurry on a fine screen that had strengthening wires spaced every few centimeters apart. These wires produced a characteristic pattern of slightly thinner paper called the watermark—in this specific case called chain lines. Actually, the watermarking is more subtle than just the chain lines. Papermakers usually added an additional wire logo to the form, thereby placing an identification mark on the sheets. In Copernicus' book, the paper contains the letter *P* on each sheet.*

Petreius printed his edition on sheets measuring 40 by 28 centimeters, a standard size known as pot paper. These dimensions have the proportion $\sqrt{2}$ to 1, which means each time the sheet was folded down the middle, the resulting page retained the same ratio of the sides, just like the metric *A* pages currently used in Europe. Petreius printed two pages side by side on the sheet, and then two more on the opposite side. When folded, the sheets produced two conjugate leaves. The technical bibliographic description of *De revolutionibus* is "a folio gathered in pairs." This means that two separate folded sheets were used, one inside the other, to make a signature of eight pages. Each signature was given a letter, and each leaf within it a consecutive number, so the leaves of a signature could be designated, for example, as *C1*, *C2*, *C3*, and *C4*. The conjugate of *C1* was *C4*, and *C2* was

---

*It would be pleasant to think that the *P* stood for the printer's name, Petreius, but *P* is such a common watermark that experts believe it simply stands for *Papier*, the German or French word for paper.

The papermaker lifts a thin layer of slurry onto the frame whose wires will produce the watermark on the sheet of paper, from Jost Amman's 1568 woodblock.

conjugate with $C3$. The P watermark appeared only once on a sheet, so only one leaf of the conjugate pair will show the symbol. Either $C1$ or $C4$ will have it, and either $C2$ or $C3$.

In Clavreuil's copy, the spacing of the chain lines of the paper was just right, but the P watermark was totally lacking, a clear giveaway for the facsimile leaves. I should hasten to point out that Clavreuil was making no secret of the problematic state of the book; his price clearly flagged the fact that she was a very sophisticated lady. By and by, the book was sold to an undisclosed buyer, so I don't know whether the current owner has any notion about the sophisticated nature of his bargain-priced but still costly first edition.

Sophistication can take various forms, not all equally benign. I vividly recall a visit to Jake Zeitlin, a famed Los Angeles dealer whose various specialties included early science. He thrust into my hand an open book (*not* a Copernicus) with the challenge, "Which side is the facsimile?"

I first examined the watermark. Holding the pages to the light, I checked the chain lines on Zeitlin's book. Both sides of the opened

The bookbinder from Jost Amman's woodblock in Hans Sachs' Eygentliche Beschreibung aller Stände *(Book of Occupations) (Frankfurt a. M., 1568).*

page spread matched. The facsimile had been made on old paper of the correct sort.

Next, I scrutinized the bite of the type, the way the individual letters pressed into the paper. It was very hard to tell which side was false, but I made a shrewd guess based mostly on very slight differences in the impression. Jake smiled and turned to the pastedown leaf inside the front cover, where he had written which pages were replacements. I had guessed right. The problem was, he had recorded the information in pencil, and there was no way to guarantee that subsequent owners would be so scrupulous. In any event, it was an educational experience for me to see how cleverly the facsimiles could be made.

Another curious case of a sophisticated copy is the *De revolutionibus* in the Dibner Library at the Smithsonian Institution's National Museum of American History. An inconspicuous marginal wormhole goes through a substantial part of the book, but it stops at the errata leaf and then continues on the other side. Obviously, the errata leaf came from elsewhere. Only about 30 percent of the copies include the errata leaf, which was

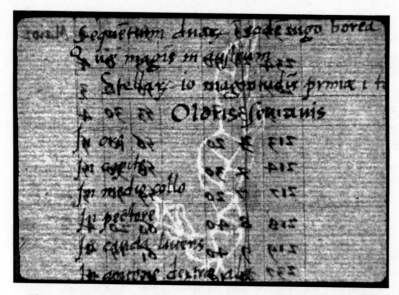

The watermark on Copernicus' manuscript in the Jagiellonian Library in Cracow. Charles Eames made the picture by backlighting the page so that the mirror image of the back side of the leaf is superimposed on that from the front side of the manuscript.

printed immediately after the regular print run, but the majority of copies apparently never got such a sheet. Purist collectors would prefer to see a copy with the elusive leaf present. Because Petreius printed the errata on whatever sheets of paper were available, the paper itself gives no guarantee regarding the authenticity of the errata leaves. I would suppose that the Dibner's errata leaf is genuine and was simply lifted from some other copy, but it is hard to be certain.

Sometimes a book passes through an auction with its faults clearly mentioned, only to reappear for sale with nothing missing. In 1979 a second-edition *De revolutionibus* clearly specified as incomplete came under the hammer at Van Gendt, a major auction house in Amsterdam, and was purchased by Librarie Thomas-Scheler in Paris. Subsequently, it was listed in a bookseller's catalog from Geneva, with no mention of any defects. I asked Bernard Clavreuil what had happened, and he replied unabashedly that he had commissioned two facsimile leaves, and he even gave me photocopies to show how good they were. He added that the restored book had gone into the collection of P. Z., but he didn't reveal who P. Z. was. So there was another challenging mystery for me.

Who was P. Z.? After making some inquiries in Paris, I had my suspicions, but for ten years I couldn't be sure.

Then, in 1995, I got a call from a prominent New York dealer. "We've just bought a second edition at a Paris auction, and I've got a question about the final leaf with the printer's anvil emblem," he said. "It's wrong-way around and looks fuzzy and fake."

I inquired what sale was involved, for it had been hard for me to keep track of the Drouot auctions because it is not a unified auction house like Sotheby's or Christie's, but rather, an umbrella organization for a conglomerate of dealers. The sale of Phillipe Zoumeroff's collection, he informed me.

"Jackpot!" I thought, and then I broke the news. "Are you aware that two leaves are in facsimile?"

There was a stunned silence at the other end of the phone. "How do you know?"

"Because Bernard Clavreuil gave me copies of the facsimiles."

That information undid the sale. The book was returned, and eventually Drouot offered it again, noting that the pages had been added but without a clue that they were actually facsimiles. It is only fair to add that eventually I acquired the copy, took out the two facsimile leaves, and replaced them with genuine leaves from a broken copy I still own. As for the anvil leaf, I'm sure it's genuine, but like most of the copies of that edition, it has the printer's mark very fuzzily printed. In 1999 I sent the book to the Reiss auction in Germany, where it fetched DM 26,000 plus the 15 percent hammer fee, and since its history is fully documented in my *Census*, there is no secret about its checkered past.

I FOUND THE wildest case of sophistication, if it can be called that, in one of the most unexpected locales. The trail that eventually led to a copy in the Victoria and Albert Museum started in 1976 when I investigated a printed reference work called *Book-Prices Current*, whose annual volumes list the rare books sold in London in that year. In one of the early yearbooks I found a reference to a first-edition *De revolutionibus* offered at the 11 November 1897 auction by Sotheby, Wilkinson, and Hodge under the heading "A collection of books formed by an amateur." The book was described as having a richly tooled binding from the famous sixteenth-century French collector Jean Grolier. I perked up when I saw this entry, because I hadn't by then found any copy with a Grolier binding. However, the auction description went on to say, "Note—the books in this library were more remarkable for their binding than for their rarity. The bindings, though modern, were historically correct. They were, in fact, modern imitations of very high quality." In other words, fakes.

Grolier formed such a distinguished library, in fact, that the book collectors' club in New York City is named after him; the Grolier Club is a sufficiently major institution to have its own librarian. I figured the club must keep track of all the known Grolier bindings, so I dropped them an inquiry. Yes, came the reply, they could tell me where the copy of Copernicus' book was if it were genuine, but as for fakes, I should ask Howard Nixon, former keeper of rare books at the British Museum.

I still treasure the reply from Nixon, who was then librarian at Westminster Abbey. "It is very nice to be asked a question that one can answer so easily," he wrote, and he told me that the binding reposed safely at the Victoria and Albert Museum in South Kensington, London. The binding was one of a number made by the French forger Louis Hagué. Many of these forgeries had been sold by the reputable London firm of Bernard Quaritch to a nineteenth-century English collector, John Blacker. Even after Quaritch became suspicious of the bindings, Blacker refused to believe they were false.

In due time I made the pilgrimage to the V & A, a magnificent museum of decorative arts of all ages. The Copernicus volume, too, was a showy sample of decorative arts, its splendid binding as fake as a three-dollar bill, and its genuine old bookplates equally fraudulent in their placement in this particular copy. At least there were no anonymous annotations whose source might have been destroyed by the nineteenth-century rebinding scam.

The V & A copy is unique in the census with its deliberate fake binding. Many copies have been rebound, of course, to replace shabby or tattered covers, without any intention to fool anyone. On the market, copies in original bindings in good condition fetch a higher price than those rebound, so an owner always faces a dilemma in replacing an old but dilapidated binding. Occasionally I have encountered disgusting cases of modern library-buckram, but more frequently tasteful imitations of earlier binding styles, though never undertaken with the goal to mislead a naive collector. There is a sort of intermediate situation, however, when an old binding has been transferred from another book. Cognoscenti call this a *remboitage*,* and when it is recognized, it definitely lowers the value of the book compared to a copy with a comparable original binding.

WITH SO MUCH indirect experience with book auctions, I finally decided I should see one for myself. So after the manuscript of the *Census* had been sent off for publication, I resolved to go to a Sotheby's auction to observe a *De revolutionibus* on the block. Accordingly, early on the morning of 16 November 2001 I flew to New York for the sale of the Friedman collection.

The collector, Meyer Friedman of San Francisco, was the medical doctor who made famous the concept of Type A and Type B personalities. He demonstrated not only that certain personality traits correlated with higher risk for heart attacks, but that by working to suppress some of the pushy or aggressive behavior, Type A persons could actually reduce their chances of having a heart attack. As a collector Dr. Friedman acquired the great medical titles, such as Andreas Vesalius' *De humani corporis fabrica* (1543), the great illustrated anatomy text, and William Harvey's much rarer *De motu cordis* (1628), the first correct analysis of

---

*A French word meaning more generally something placed into a socket.

The fake Grolier binding on the De revolutionibus *in the Victoria and Albert Museum in London.*

the circulation of the blood. But his collection also included pure science works as well, including those by Copernicus, Newton, and Einstein.

I never met Dr. Friedman because I had already seen his copy of Copernicus' book when it was still in the hands of its previous owner, Marcel Chatillon, a French surgeon. Chatillon practiced in the former French colony of Guadeloupe, and he collected regional art and Americana. *De revolutionibus* fell in the latter category because Copernicus mentioned America, "named after the sea captain who discovered it." The Polish astronomer was misled on this matter because he relied on Martin Waldseemüller's *Cosmographia introductio* of 1509 for news about the New World, but nevertheless his fleeting reference to Amerigo Vespucci was enough to qualify his book as an Americanum.

Dr. Chatillon showed me his copy in Paris. I still cringe whenever I recall our conversation: Chatillon spoke no English, and my spoken French could charitably be called rudimentary. Nevertheless, I heroically carried on in French as I examined his relatively ordinary copy, measuring its page size, noting the binding and the position of the errata leaf, and hastily scanning for annotations. As it turns out, I missed the two minor marginal notes and failed to investigate carefully the most interesting point in the book, its somewhat weathered final leaf.

In 1978 Dr. Chatillon sold his Copernicus copy, and it went into the Friedman collection. When Dr. Friedman acquired it, a first edition was still relatively easy to find, and he got the book for $65,000. During the 1970s and 1980s twenty copies passed through the market, but ten of them moved to institutional libraries, gradually contracting the number of copies available for private owners. By 2001, when Sotheby's announced the Meyer Friedman sale, only about twenty copies remained in private collections, and with the rise in cyber fortunes owned by the newly wealthy with interests in science and technology, the competition to obtain a copy had built to volcanic proportions.

The austerely elegant auction room at Sotheby's took me by surprise. There was not a book in sight apart from the sale catalogs carried by the potential buyers, a group of twenty-five who were mostly dealers. I recognized several from London and others from the West Coast. On a raised dais along the side and front were nine phone stations for anonymous bidders, and in the back was a long table with computers, whose operators apparently controlled a screen in the front that would show the lot number and current bid in ten different currencies. As the group assembled, I introduced myself to Selby Kiffer, one of the Sotheby experts, who promptly asked if I wanted to see the Copernicus. Indeed I did, so he fetched it from the back room.

Several dealers gathered around as I reacquainted myself with the book. I took a hard look at the last leaf, a bit scruffy with a library stamp totally removed and carefully patched with a paper repair. Clearly, the leaf had been remounted, but I saw no reason to doubt its legitimacy in the volume. I turned the book back to Kiffer and settled down in one of the hundred portable chairs whose rows formed a square in the middle of the room. The clerks took their places behind the phones. David Redden, the auctioneer, ascended the spotlighted podium in the front and asked, "Shall we begin?"

The bidding on various lots moved sequentially and swiftly, by standard increments, $25 for the lower range, for example, or $500 when the bids got over $5,000. There was no accidental bidding by someone scratching his nose—even the catalog made that explicit. Bidders held up their hands, and Redden frequently indicated who had the current bid, with a phrase such as "in the third row" or "Selby's phone." Otherwise the auctioneer limited himself to the figures—the lot numbers, which came at a rate of about one a minute, and the bids. It was very rare that he announced the author or title on the block.

By the time the sale approached lot number 34, the *De revolutionibus*, the assembled dealers realized, with some shock and even more curiosity, that about 80 percent of the books were being snatched up by an anonymous telephone bidder, designated as L020, and most of the others were going, after a duel of phones, to a second anonymous bidder, L029. Several of the rarer items had already gone well above the high end of the estimated bids.

The auction had been under way barely half an hour when lot number 34 flashed on the screen. I set my stopwatch. Did the bidding start at

$100,000 or $200,000? I was too excited to remember, but the auctioneers say it was $150,000. I was later sorry that I hadn't registered for one of the numbered bidding paddles so that I could have made the first offer, even though I would have had to mortgage the house to back it up. The bids came from the floor at a furious rate, notching up by $25,000 intervals. At first the bids came too rapidly for the phone bank to compete, until the offers passed the high estimate of $400,000—elapsed time, forty-five seconds. The highest floor bid came in at $550,000, already past the previous auction record for a first edition of *De revolutionibus*, and then the battle devolved to three anonymous phones. I could overhear a Sotheby's representative say, "It's now at $600,000. Do you want to raise?" And L020 raised.

Moments later, the bidding stopped. The auctioneer tapped the hammer at $675,000. With Sotheby's hammer fee, the book had cost L020 almost exactly $750,000. I clicked my stopwatch. The bidding had taken two minutes and sixteen seconds.

The sense of curiosity in the room was palpable. Who were these mystery bidders? The rest of the auction gave only limited clues. The initial printing of a thousand copies of Darwin's *On the Origin of Species* had sold out on the first day, but the edition is not particularly rare and Sotheby's high estimate of $35,000 was entirely reasonable. Nevertheless, the two telephone duelists drove the price to a sublimely ridiculous $150,000. Experienced collectors would have already owned this classic, and no institution could have been so reckless with its funds. Clearly, the two collectors were beginners, going after an instant library, keenly chasing both the medical and the more strictly scientific items.

In the auction's aftermath there was a great deal of speculation, and no hard facts, about the identity of the well-heeled buyers. And there was a lot of second-guessing about the status of the Copernicus book itself. "There is a real problem about that last leaf," Rick Watson assured me. A longtime and well-trusted friend among the London book dealers, Rick had noticed that the chain lines on the leaf did not run parallel with the edges of the printing, a detail that had escaped my inspection. Since the conjugate leaf was not skewed in the printing, the final leaf had to have come from somewhere else. Clearly, it was a sophisticated copy.

I checked my records again. The book had been auctioned at Drouot in 1973, and sure enough, it lacked the last leaf. Had it passed through Scheler's shop? The team that added a fake water stain in the Cal Tech copy could have fabricated a page with a fake patch to cover a phony library stamp. That would be enough to keep almost anyone from looking too hard at the page. To my eyes the leaf passed muster; it was similar to a number of copies I had seen in which the wear and tear on the first and last pages was relatively harsh. My notes showed that the Scheler shop formerly had two copies in which the last leaf was probably in facsimile. The genuine leaf in the Chatillon-Friedman copy could have come from one of those Parisian copies. Someday, no doubt, their owners' names will become known. Then I would certainly relish the chance to inspect any of those copies more thoroughly.

Meanwhile, there are several collectors impatiently waiting to spend a million dollars on an unsophisticated first edition of Copernicus' book in an original sixteenth-century binding—preferably a copy with a known provenance and clear title.

Chapter 14

THE IRON CURTAIN:
BEFORE AND AFTER

IN LENINGRAD a gentle snow was falling even though the calendar read April. From my window in the old Astoria Hotel I could look out on St. Isaac's Cathedral. A woman was sweeping the snow off the sidewalks. A Russian historian of astronomy, Nina Nevskaya, had come to meet me at the airport, flowers in hand, but she hadn't been allowed to ride back into town with me in the car Intourist had provided. I was an alien speaking a foreign tongue in a strange and forbidding land. It was 1976, Leonid Brezhnev was in power, and Leningrad was very much behind the iron curtain.

My first goal was not Copernicus but the world's greatest concentration of Kepler manuscripts. This lion's share legacy had been in St. Petersburg ever since the prolific mathematician Leonhard Euler had persuaded Catherine the Great's minister of finance to buy them for the Russian Academy of Sciences in 1773. I was staying within walking distance of the Academy Archives, where the eighteen thick volumes of Kepler's manuscripts are now preserved, but unfortunately the demon Intourist driver had whisked me off from the airport before I could get good instructions from Nevskaya as to exactly where the archives were. On Thursday morning I walked across the Neva River, past the Lomonosov Museum into the thicket of streets on the island where the scattered academy buildings are located, and promptly got lost. I ended up at the Academy Library, which I had visited before in order to see its Copernicus books, and there I encountered an American graduate student who shepherded me down the street to the archives. Once inside, I found it exciting to handle, among others, the volume of Kepler's Mars researches, whose microfilm I had pored over so intensely.

Eventually, I ran up against some mysterious fetching limit. Had I had my quota of volumes for one day? It's like being an exile at the end of the Earth when you can't speak the language. In any event, Nina Nevskaya found me and took me in hand; we went by trolley to the Saltykov-Shchedrin Public Library to begin the paperwork so that I could see its Copernicus volumes later. Friday was taken up at the Pulkovo Observatory on the outskirts of Leningrad. There I gave a colloquium, speaking as distinctly as I could in English and *very* slowly. Afterward I had a chance to see its venerable library. The "great" works—the Keplers and Copernicuses—had seemingly all been transferred to the Academy Library, which left a host of minor works, one of the most complete and fascinating collections of little sixteenth-century textbooks I have ever examined.

This left Saturday to see the Copernicus books. Nevskaya fetched me at the hotel—she was afraid I would get lost again—and by 11 A.M. I was back at the Academy Library and had before me its three *De revolutionibus* copies. As I knew from a previous visit, each of them was annotated. Nevskaya left me there so that I could work at my own pace in documenting the marginalia more carefully than I had done six years before. The first copy, a first edition, had an English provenance and had been bought in London by a Polish aristocrat; precisely how it got from his Warsaw library into Russia is unknown. The main annotations, in a fine, nineteenth-century hand, derive from a copy of the second edition now in the Warsaw University Library. Another first edition contained typical and rather uninteresting sixteenth-century student notes derived from the text itself; at the end of the volume there was a long account of what the Egyptians knew of the sky and geometry (in truth, not very much). The second edition was the most interesting of the trio, with a relatively large number of references to sixteenth-century textbooks and, on its title page, references to two scriptural passages against the mobility of the Earth and a quotation from Ptolemy indicating how ridiculous the Earth's motion would be.

For better or worse, I was completely on my own getting back to the Saltykov-Shchedrin library. I returned to my hotel and lunched on chocolate and cheese that I had brought from Sweden on the previous segment of my expedition. That was easier than coping with the language barrier. Then I walked back down the Nevsky Prospect to the library. Before the revolution the Saltykov-Shchedrin was the great westward-looking library in Russia, with the kind of fabulous collections associated with major national libraries, but it had been reduced to secondary status when the Lenin State Library in Moscow took the lead. Even now the Saltykov-Shchedrin is a major depository of early books. I knew its rare book library was closed on Saturday, so it seemed inconceivable that I would get anywhere on my own in the main reading room at 3 P.M. Saturday afternoon. Nevertheless, I went through the motions: queuing up to check my coat, showing my card and getting a control pass, and then wending my way through the labyrinthine corridors to the reading room, exactly as in the dry run made with Nevskaya on Thursday. I handed over my control pass without saying a word. The attendant beamed and promptly turned up with two Copernicus books! They had been transferred from the rare book room to the main reading room so that I could inspect them.

I had guessed the library might have a copy of the second edition, which it did, but this copy was pretty uninteresting. Its copy of the first edition was another matter. When I turned to the title page, the handwritten motto "*Axioma astronomicum . . .*" popped right out. It was the famous—to me at least—motto that Reinhold had inscribed in his book, and which Wittich had transcribed into his *De revolutionibus* that is now in the Vatican— though then I was still under the misapprehension that Tycho had annotated that one. I quickly paged through the Leningrad copy, noting the standard Reinhold passages copied into the margins. But there was yet another hand with "*Ego (Tycho Brahe) . . .*" in three places. These first-person references seemed at least superficially to match what I then mistakenly believed was Tycho's hand. I felt almost giddy two hours later as I walked back to the Astoria, confident that I had bagged yet another *De revolutionibus* owned by Tycho Brahe.

Sunday was a day for the museums. One of my rules of Copernicus chasing was to arrange the itinerary so as to be in an interesting place on Sunday, since opportunities to look at books on the Sabbath are decidedly unusual.

I walked myself limp in the Hermitage and in the afternoon visited the Museum of the History of Religion and Atheism in the former Kazan Cathedral. Copernicus, Galileo, and Bruno loomed large there, but the place of honor was reserved for wax models of an inquisitorial torture chamber à la Madame Tussaud.

Early Monday morning I took the black case with my photographic lamps to the archives where the Kepler manuscripts resided. The one assistant who spoke some English looked at it in horror.

"I will ask our director," she said. Moments later I was ushered into Madame Director's office, given a desk, and told to go to work. All day long I photographed Kepler manuscripts to my heart's content, making a series of color slides that have decorated my lectures ever since.

Elsewhere I had set things in motion to photograph the *De revolutionibus* at the Saltykov-Shchedrin Library, but there the answer was "nyet." I had spoken too enthusiastically about my discovery, and the Russians decided they wanted to publish it themselves. By the time they did, I had found out that the handwriting was not Tycho's, so their account was pretty well botched up. The book still remains one of the mysteries of the census, but it clearly reflects some lost copy with Tycho's own annotations.

Meanwhile the assistant in the archives had got into a dither worrying about how I could get undeveloped films out of the Soviet Union. Even more troubling were the undeveloped rolls I had brought in from Cairo and Uppsala, two places I had visited en route to the USSR. I spent a rather sleepless night just before my departure. Nevskaya and the assistant at the archives prepared an official document for me and took me to the airport in an academy car. In the end the customs officer looked rather bored by it all and didn't open anything.

ONE DAY AS I was casually looking at slides on my light table, at images I had made of early books, three manuscript initials suddenly riveted my attention. There they were, *EWL*, in fancy script on the title page of the 1515 *Almanach nova*, a book of planetary positions. What jolted my memory was the fact that the same fancy unidentified initials appear on the title page of the *De revolutionibus* owned and annotated by Johannes

Kepler. We know Kepler was devastated when his library was temporarily locked up during the Catholic Counter-Reformation, but we have precious little information about what books he had in it because almost nothing from his library has been identified. Was I just about to identify a long-lost book once owned by the German astronomer?

Of course, merely the initials *EWL* were pretty slim evidence. But as I took a close look with a magnifier, I recognized as well the signature of another owner, Hieronymus (or Jerome) Schreiber, and that name, too, was on the title page of Kepler's copy of Copernicus' book. It seemed highly likely that the volume in the slide had also belonged to Kepler, who seldom inscribed his own books. I had photographed it in the university library in Wrocław—the former German Breslau—and I immediately wondered whether other volumes there might have come from Kepler's library, which seems to have vanished with scarcely a trace.

Because of the Copernican festivities in 1973 and their aftermath, I had traveled to Warsaw fairly frequently over the years and had learned the ropes well enough to cope even on the rare occasion when by some snafu no one came to meet me at the airport. After General Wojciech Jaruzelski instituted a military government late in 1981 to quell dissent from the ever stronger Solidarity movement, I was no longer keen to return to Poland, but I remained very curious about the books in Wrocław and concerned about my colleagues there. So in 1984 I decided to go once again to Poland and to revisit Wrocław.

In the university library I spent the better part of a day writing out call slips for every sixteenth-century astronomy title I could think of that might have been owned by Kepler. One minor detail bothered me somewhat: I couldn't find in the catalog a card for Rheticus' *Narratio prima*, published three years before *De revolutionibus* and containing the first printed report of the new heliocentric cosmology. A few months before, the only known copy still in private hands had been auctioned in New York for $400,000. Not unexpectedly, such an astronomical price had the effect of bringing another copy out of the woodwork. Friends in the book trade asked if I knew anything about a library called "the house of Mary Magdalene," because a copy that had mysteriously turned up in Paris included a Latin stamp, "Aed Maria Magd." We could only conclude that it was some obsolete establishment.

Since I had seen the *Narratio prima* on my previous visit to Wrocław, I still had its call number, so I included it on the list of books to be fetched even though I hadn't been able to find the card in the catalog.

After some hours I had before me a huge array of books, about a hundred titles. First the good news: Hidden deep inside the *Almanach nova* with the initials *EWL* and Schreiber's signature, I found a whole manuscript page in Kepler's hand, previously unknown.* My hunch had been right; the volume of planetary positions had belonged to Kepler. The bad news: There were no other books with a Keplerian provenance. However, I kept finding book after book with the stamp "Aed Maria Magd." And most disconcertingly, the *Narratio prima* seemed not to be included among the cart full of volumes that had been fetched for me. As I left for lunch, I drew the librarian's attention to the fact that Rheticus' book had been overlooked.

When I returned in the afternoon, the librarian's face was ashen. The *Sammelband*—that is, a collection of small books bound together—supposedly containing the *Narratio prima* had in fact been fetched for me, but Rheticus' small tract was missing, carefully removed from the collection, and its loss cleverly concealed by pressing the book more tightly together. In my absence the worried librarian had gone back through the records to the previous time the volume had been fetched, and then he had collected all the other books that had been called for at that same time. Two other small tracts turned out to be missing as well.

"Let me guess," I said. "One is Copernicus' small book on trigonometry, *De lateribus*, and the other is Rheticus' *Orationes duae*."

The librarian's astonishment was palpable. How could I have known? "Easy," I replied. "Those are the other two unusual books also being offered in Paris. You had better get in touch with Interpol right away." Obviously, the thief had been knowledgeable about old books and clever in the way he covered his tracks by removing the cards from the catalog.

---

*Or so I thought. I later discovered that the redoubtable astronomical bibliographer Ernst Zinner had noticed the Kepler manuscript but had not followed up on it.

It was much easier for me to give this advice than for the librarian to take it, for at that time no academic wanted anything to do with the bureaucracy of the Polish military government. Besides, the librarian himself could get in trouble for not having proper security in place. If truth be told, this is the gut reaction of librarians worldwide, who tend to keep their losses dark secrets.

After I returned home and a few months had passed, I realized that nothing was happening on the international front with respect to the missing books. By chance I encountered the president of the Polish Academy of Sciences in Harvard Yard, and I told him what I knew about the theft. After that, the authorities sprang into action, and Interpol took an active interest in the matter, both in France and in the United States. On behalf of Interpol, the Boston office of the FBI sent an agent to interview me, and I told him what I had heard about the books being offered for sale in Paris.

The next thing I learned was that a lawyer had walked into the Polish embassy in Paris, had plunked down a package saying, "My client no longer needs this," and had turned on his heels and left. The embassy was totally baffled to find two old books in the package, but shipped them to the History of Science Institute in Warsaw, where my friend Jerzy Dobrzycki was about to become director. He and his colleagues recognized the contents, so they returned the *Narratio prima* and the *De lateribus* to Wrocław.

And what about the third missing book, Rheticus' *Orationes duae*? It had been sold to a dealer in Liechtenstein, who had in turn sold it to a library in Rheticus' birthplace in Austria. The Liechtenstein dealer stonewalled, and the Wrocław library did not have enough funds to force an international lawsuit, so the stolen copy of the *Orationes duae* has not come home.

THERE WAS always a certain amount of tension involved with going behind the iron curtain, sometimes with unexpected tragicomic elements. Miriam and I particularly recall our last visit to Halle, the successor university to Wittenberg. In an earlier visit to this East German town I had uncovered the fact that the second-edition *De revolutionibus* listed in the library's catalog had disappeared. Over the next few years I made two interesting discoveries relating to Halle's collection. First, I found its lost *De revolutionibus* in the Lenin State Library in Moscow, inscribed by the university rector as a gift to the Russian general who had "liberated" them. Second, the Halle University Library apparently had another copy of Copernicus' book, which I was determined to see.

On the second visit, in 1987, my way was temporarily barred by a party functionary who worked as an assistant librarian. He proved to be a formidable gatekeeper, not at all ready to give me access to the library's precious volume. He insisted on giving me a lecture, in German, on communism, adding that some of the older professors still believed in God, but the younger ones were more enlightened. "What do you need to know about the book that you can't learn if I look at the book and tell you?" he wanted to know. In the strained conversation, which stretched my spoken German to the limit, I finally caught on that he was highly resistant to showing me the book because one of the library's copies was missing.

"Ah, but I know where your missing book is," I announced in undeclined German.

Now the Marxist bulldog perked up. Where did I think it was? he wanted to know. I explained that I had seen it in the Lenin State Library and that it was the gift of the university rector to the liberating Soviet army.

"*Ausgeschlossen!*" he replied. "Completely out of the question. The rector wouldn't have had any right to give the book away without the approval of the Senate. You are mistaken."

"It's in Moscow," I retorted, "and it's bound with Stadius' *Tabulae Bergenses.*"

With that he snorted and bounded out of the office to have a look at the card catalog to see if the missing copy was bound with a book by Johannes Stadius. Presently he returned, completely shaken. It was as if his confidence in the entire system had just collapsed. The catalog had confirmed that the missing Copernicus volume was indeed bound with the Stadius, so I could hardly have invented the report. Shattered, he led us into the reading room and showed me the remaining copy of *De revolutionibus*, meanwhile breaking every library rule that he had earlier forced me to read. He leaned on Copernicus' book, took notes with his pen, and spoke English. Behind his back the other librarians winked at us. My wife, Miriam, the silent observer of this episode, could hardly contain her amusement, especially as she watched the expressions on the faces of the other staff members.

Later I got an apologetic letter from the head librarian, who hadn't realized we were visiting. I have often wondered what became of that assistant after the demolition of the Berlin Wall in November of 1989.

The reunification of Germany brought one pleasant if unanticipated benefit for the library in Zittau, a small East German border town tucked in the corner of Germany, Poland, and Czechoslovakia. I suppose few Americans had ever penetrated to the Christian-Weise-Bibliothek, so I was treated like a celebrity when I arrived there in August of 1976. The staff showed me not only their first-edition *De revolutionibus* but the original printing of Newton's *Principia* and a series of other scientific rarities including an almost complete set of major Keplerian titles. It was worrisome to see so many treasures in a relatively unpretentious and insecure library, and my apprehensions proved all too well founded thirteen years later.

Late in the summer of 1989 the London book dealer Rick Watson called to see if I could shed any light on the provenance of a *De revolutionibus* to be offered at an auction in Cologne. He told me that it contained an inscription dated 1733, and also that it had an errata leaf inserted between folios 194 and 195. This latter information made the search quite easy, for I had kept a separate list of copies with the scarce errata leaf, and I could almost instantly establish that the only copy known to me with an errata leaf bound in that spot was the copy recorded in Zittau. My notes also showed that this copy had an inscription dated 1733. Clearly, the copy at the Cologne auction had come from Zittau, but the question remained whether the book was stolen or legitimately deaccessioned.

I attempted to contact my East German colleagues in Jena, but as luck (or a collapsing infrastructure) would have it, the university's fax was out of order. Next I tried to telephone the director of the Archenhold Observatory in East Berlin, a friend with a deep interest in the history of astronomy, but a regional telephone operator informed us that no one answered there. Fortunately, the German astrophysicist who had assisted with the call remembered that an East German astronomer, Hans Haubold, was posted to the United Nations in New York, so we contacted him.

"Our fax does work," Haubold assured us, so he promised to investigate the situation at Zittau. Meanwhile, the auction date was rapidly approaching, and we heard nothing from him. At last he called to say that Zittau owned the book, but he hadn't been able to ascertain if the book was actually there.

"Of course it isn't," I explained again. "It's in Cologne, and if the library at Zittau isn't trying to sell the volume, they had better put in a claim instantly."

Rick Watson agreed to break the news to the auction house, which he did, but then I lost track of the events. Later I got two mutually contradictory reports of what happened. In the first version, the East German ambassador had put in a claim for the Copernicus book, but something about West German law required that if a book had been advertised for auction, as it had been in Cologne, then it couldn't be withdrawn. According to the West German press, the auctioneer had stated that the ownership of the *De revolutionibus* was in question, and then proceeded to auction the book to a dealer not noted for his scruples.

Two months later I received a letter from the East German plenipotentiary to the United Nations, stating quite clearly that the request from the German Democratic Republic for assistance had been turned down by the Federal Republic of Germany. According to this second version, the East Germans then had hired a lawyer, and the book was seized by the Cologne district attorney two days before the auction. However, the book had not yet been returned. In a divided Germany the Zittau library did not have the hard currency to wage a legal battle to get the stolen book back. The ambassador included his thanks, but it wasn't obvious that the book would be returned to Zittau.

When the Berlin Wall came tumbling down, the legal situation quickly changed. Finally, early in 1992, a letter arrived from Zittau, thanking me for my intervention, and reporting that as of November 1991, the book was safely back in the Christian-Weise-Bibliothek. "This is the second time we have temporarily lost the book," the librarian added. "During the Second World War it, together with many other valuable books, was taken into Czechoslovakia, and they did not come back until 1957."

SOON AFTER the Berlin Wall came down in 1989, the Soviet Union crumbled, opening a window on a closely guarded secret. I had accidentally learned—an unintentional mistake on a Soviet librarian's part, as it turned out—that the Lenin State Library in Moscow held six copies of Copernicus' book, three of the first edition and three of the second. She had shown all six copies to a visitor, who had noticed which ones were annotated. I had seen five of them, but try as I might, I could never examine the sixth copy, one of the three second editions. I was told it was "in conservation" and unavailable, and each time I requested a microfilm, the library sent me a reel with a copy I had already inspected. Finally, Miriam said, "Look at this article where an American scholar describes all the excuses the Russians give when they don't want you to see something. It's the same thing that is happening to you!"

So I had to face the possibility that my census would simply have to omit the description of a copy known to be rather heavily annotated. There the matter stood until the winter of 1990–91. Then, a year after the wall fell, and as the Soviet empire was collapsing, I got a telegram from a historian of astronomy in Moscow. "If you want to see the sixth Copernicus, come now. You will be our guest while in Russia."

Extravagant as it seemed to fly to Moscow in January to inspect one copy of Copernicus' book, it also appeared to be an especially interesting time to see what was going on in the former Soviet Union. Within twenty-four hours of my arrival in Moscow I had seen and photographed the Copernicus volume, a particularly interesting second edition well annotated by Herwart von Hohenburg, the chancellor of Bavaria and a frequent correspondent with Johannes Kepler (plate 7c). But one thing struck me as odd: Clearly, the book had never needed, nor ever had, any conservation, despite the librarians' previous excuses when I had asked to see it. When the young postdoc who had taken me to the library remarked that he had been there the previous day to make sure the book would be available, and that he had been quizzed as to whether I *really* had permission to see the book, I realized that something fishy was going on. I finally confronted my host, who with mild embarrassment revealed the real situation.

After the Second World War the Soviets captured truckloads of books from East German libraries in reparation for their own tremendous losses. However, by the Geneva Convention, these books could be considered cultural treasures and hence subject to return. In particular, the Copernicus volume had come from the Leopoldiana, a venerable natural history academy in Halle. Under the circumstances the Soviets naturally didn't want foreigners mucking about in their holdings. Eventually, I had an opportunity to mention this to the head of the Library of Congress, a specialist in Russian studies, who allowed that the experts had always suspected this, but my report was the first actual evidence.

AS THE IRON curtain disintegrated, it occurred to me that I had never really examined one of the important collections that was, in a sense, a distant background to my Copernicus chase story. The Crawford Library at the Royal Observatory, Edinburgh, where I had stumbled across the well-annotated *De revolutionibus* owned by Erasmus Reinhold, had been inspired by an extraordinary Russian collection. The Edinburgh library was established by a scion of a noble Scottish book-collecting family, the Lindsays. Alexander William Lindsay, the 25th Earl of Crawford and 8th Earl of Balcarres, had formed the leading private European library in the second half of the nineteenth century. It included an outstanding collection of Bibles in all languages, among them the Gutenberg Bible, all the early editions of Luther's translations, and the exceedingly rare Eliot Indian Bible printed in Cambridge, Massachusetts, in 1663. Natural history, the classics, and manuscripts in oriental languages all had their place as well. The earl's eldest son, James Ludovic Lindsay, developed a fascination with astronomy, and his well-equipped observatory became the envy of the professional astronomers throughout the land. In such a home environment he could hardly help but include a collection of rare astronomy books in his observatory, and his astute buying built up a fabulous scientific library. He donated this collection, along with his instruments, to Scotland in 1888 when his tastes turned increasingly to his yacht and his stamp collection.*

James Ludovic Lindsay, who became the 26th Earl of Crawford and 9th Earl of Balcarres, modeled his collection on the Pulkovo Observatory's library, which is situated just to the south of St. Petersburg. That library boasted a magnificent collection of early astronomical titles, and what still remains the world's finest assembly of comet tracts, despite Lord Crawford's energetic efforts to catch up. Crawford must have been especially pleased when Otto Struve, director of the Russian observatory, paid a visit in 1875 and brought along as a gift an autograph leaf from Johannes Kepler's legacy, then housed at Pulkovo.

In 1994, when the Scottish National Museums put on a special exhibition from the Crawford Collection, I had occasion to reflect on several great astronomy collections, in each of which I had worked through the shelves, book by book. "It is hard to rank one above another," I wrote in the exhibition catalog. "Each is astonishing in its own way, with bibliographical discoveries awaiting diligent scholars." Yet, as I thought about them, I realized that I had never looked carefully at the Pulkovo Observatory's library, the great collection whose published catalog had served as the prototype for Lord Crawford's own acquisitions.

I knew that the Pulkovo library had been on the front lines in the siege of Leningrad from 1941 to 1943 and that the observatory had been ruined. Of the rare astronomy books, nearly a quarter were lost. A decade after this tragic reduction of the library, Professor Alexandre N. Deutsch, who was at the time the acting director, briefly described the event.

---

*No other person has ever been a leading officer of both of the Royal Astronomical Society and the Philatelic Society!

The observatory presented an awful sight when we climbed the hill along the avenue of the park. . . . A pale, waning moon shining through a cloudy haze illuminated the unrecognizable walls of the main building, which now had holes instead of windows. The domes had collapsed and burned, as had the roof. . . . We descended into the clock basement, where the books of the Pulkovo Library had been placed at a depth of five meters. In the basement's central section the vaults were whole, but the books lay in chaotic confusion. With difficulty we searched out the box with the rare books and incunabula. Soldiers carried the box by hand to a truck at the foot of the hill. Over the following two nights the Leningrad Soviet organized expeditions of the women's militia using several trucks to carry away the surviving stocks of the library. Many staff members took voluntary part in these expeditions. At one point people were forced [by the artillery fire] to get out of the trucks and lie down in a ditch beside the highway. Fortunately the enemy shells fell on the other side of the road and no one was hurt.

I knew that the precious Kepler manuscripts, the heart of Kepler's material legacy, had been saved, but what else was left in this once-proud library? Consequently, I resolved to go to St. Petersburg to see for myself. When an old friend, Viktor Abalakin, who in the meantime had become director of the Pulkovo Observatory, learned of my interest, he urged me to come. Thus in 1996 I arranged a summer trip to St. Petersburg. Accompanying me was James Voelkel, a young scholar specializing in Kepler; he was keen to view the manuscripts I had seen, which had fortunately been moved from the observatory to the Archives of the Russian Academy of Sciences shortly before World War II.

Abalakin met us at the St. Petersburg airport, whisked us through customs, and took us to the observatory, which sits on an attractive ridge in an extensive wooded tract just south of the airport. Rather uncommonly for an observatory, Pulkovo has its own hotel, where we were soon situated in spartan but quite adequate accommodations. Fortunately, a cosmology

Comet Hale-Bopp in the sky over Pulkovo Observatory.

conference was going on, so catered meals were available with a congenial group of international astrophysicists. The historical central observatory building was a pleasant hike up a gentle slope and through the woods. It had been entirely rebuilt along the pattern of the original observatory constructed by F. G. W. Struve in 1839.

Struve, who had left Germany in 1808, worked first at the Dorpat Observatory (near Tartu in Estonia) and was later brought by Czar Nicholas I to found the new observatory. From Munich he ordered a fifteen-inch refractor, the largest in the world at that time, which he installed in the central dome. Struve also began building a first-rate library, which he cataloged in 1845 in the first volume of the observatory's *Annales*. He was especially fortunate to purchase the library of H. W. M. Olbers, the German astronomer famous for discovering the second and the fourth known asteroids, and who had formed a virtually complete collection of comet literature. Struve also purchased books especially heavily from the dealers Antiquariat Weigel in Leipzig and Bohn in London, and then he persuaded the Russian Academy of Sciences in St. Peterburg to give him the astronomical titles it had acquired through two book-collecting astronomers who had worked for the academy in the eighteenth century. The academy conceded the Kepler manuscripts as well. Struve's son, Otto, who would carry on as director from 1862, published an augmented two-volume catalog in 1860. These books constituted the core of the "Struve Library," and the printed catalogs served as the guide for Lord Crawford.

On the main floor of the rebuilt central building I found the reading room, eerily uninhabited. Locked up partly below ground level were the stacks for the old books of the observatory's Struve Library, which were promptly opened for me. A quick scan of the shelves showed that in major sections the old numbers were continuous, but in other places there were substantial gaps. Inventories had been taken in August and September 1939, on the eve of World War II, and again just after the war, so it was easy to establish which subjects had been preserved almost intact, and which had suffered heavy losses.

The Struve Library once contained an impressive showing of Kepler's printed works, twenty-four titles or editions, essentially all of his major contributions. Unfortunately, a large number of the Kepler volumes were classified in a section that was heavily damaged in the war. As a consequence, only fourteen of the Kepler titles remained. The elder Struve was particularly proud of having both volumes of Johannes Hevelius' spectacularly illustrated *Machina coelestis* (1673–79), since most of the stock of the second volume had been destroyed before distribution by a disastrous fire in Hevelius' quarters in Gdansk; the rare pair, which described the instruments and telescopes in Hevelius' Gdansk observatory, escaped the World War II bombardment. I was particularly excited to find both a first and second edition of *De revolutionibus*, especially because I had thought that the Pulkovo copies had been taken to the Academy of Sciences Library in the town and that I had already seen them there, but this wasn't the case. Both editions had been originally purchased in Wittenberg by their sixteenth-century owners, and unusually, both included the price paid, eighteen and nineteen groschen respectively, compared with the 6–10 groschen matriculation fee at the university.

On the final evening of my visit the librarian opened the stacks at night so that I could make the fullest use of my time. Since Jim Voelkel didn't have night access to the Kepler manuscripts in the Academy Archives, he came along to look at the books, and presently he discovered something I had overlooked. In cardboard file boxes on the other side of the main aisle was the famous collection of early theses and comet tracts, apparently completely intact. The comet tracts were arranged chronologically, and up through the comet of 1618 there were forty-three tracts, compared to fifty-two in the rival Crawford Collection. However, the Struve Library contained several very important ones not found in Edinburgh. I could have profitably spent a couple of more days in the comet collection alone.

Eighteen months after my reconnaissance of the Pulkovo library, bad news arrived from St. Petersburg: The precious collection had become the victim of an arson attack on the night of Wednesday morning, 5 February 1997. A gangster had broken a window and had thrown burning oily rags into the library stacks. Firemen arrived, and the books not burned were soaked with water. The news reports were somewhat contradictory, but they suggested that a thousand books were burned and the remainder badly damaged by the water.

Private communications offered the following scenario: Members of the Russian mafia coveted the beautiful and conveniently situated site of the Pulkovo Observatory for building a resort hotel and had been frustrated in their attempts to capture part of the observatory land. Three previous attempts had been made to torch the library, and an understaffed St. Petersburg police force was unable to take the matter seriously. The incunabula and some other books such as the precious first-edition *De revolutionibus* were therefore removed to a safer place. The arson took place when neither Director Abalakin nor the deputy director was in town. Tragically, those books that had managed to escape destruction during World War II became the victims of local intrigue and hooliganism.

The comet tracts in their cardboard file boxes were particularly vulnerable to fire, but apparently escaped intact. The water-soaked books were removed to the library of the Russian Academy of Sciences. In 1988 a terrible fire there had destroyed four hundred thousand books, but a legacy of that disaster is that the Biblioteka Akademia Nauk is one of the world's best centers for restoring damaged books. A carefully restored rare book does not lose its scholarly value, though it would fare less well on the rare book market.

Unfortunately, this Pulkovo tragedy has yet a further disturbing sequel. In the year following the fire, the second-edition *De revolutionibus* was stolen, probably in the confusion surrounding the arson. Its bookplate was removed and the copy brought to Munich for auction. Once again my records compiled for the census served to identify the copy, which was promptly seized by the German authorities. However, the Munich police apparently decided that since the Russians had stolen so many books from Germany, they would not send it back. So, to the best of my knowledge, the book is locked up somewhere in Germany.

At least some of the stories have a happy ending as stolen books occasionally come home. Glad or sad, they each provided terse historical lines in the Copernican *Census*.

Chapter 15

PUTTING THE
CENSUS TO BED

JUST OVER a decade after the Berlin Wall fell, and after years of anticipation, in the spring and summer of 2001 *An Annotated Census of Copernicus' De revolutionibus (Nuremberg, 1543 and Basel, 1566)* was finally in
its closing stages. The map of Europe had changed a great deal from the
time I began the Great Copernicus Chase. In my lists I had arranged the
books alphabetically by country and city, and I had to reshuffle the entries several times. When I started my quest, Europe was neatly divided
between East and West, with a line and prisonlike wall meandering
through Germany. With the demise of the wall, I had to realphabetize the
East and West German lists, merging them into a single roll. That almost
made the combined Germany the top contender for copies of the first
edition, with forty-five, running a close second to the fifty in the United
States. Within the individual countries the towns were listed alphabetically as well, and even this listing got rearranged when Leningrad re-
verted to St. Petersburg.

Eventually, I typeset the entries for Russia, which had moved forward
from its original USSR position when the census started. Next came
Spain and Sweden, and then Switzerland with a disproportionately large
section because the description of the important annotations by Kepler's
teacher, Michael Maestlin, in the *De revolutionibus* in Schaffhausen required ten pages. Throughout the entire research process for the census I
had collected photographs of critical pages to document different handwritings, for comparison of marginal diagrams, and to give the general flavor of the annotations. In the end I included a couple of these images

simply to show what nearly indecipherable hands were involved. I selected two from Maestlin's copy in Schaffhausen just to demonstrate how microscopic his handwriting was. I suspect he was very nearsighted.

At last I came to "United Kingdom (England)," another long section because it included thirty-two first editions and thirty-eight of the second, and then to one of the most interesting sequences in the entire *Census*, "United Kingdom (Scotland)." I'm still amazed by the roulette that brought so many of the most important copies of *De revolutionibus* to Scotland. This part of the *Census* begins with Aberdeen, where one of its three copies contains, on interleaves bound into the book, one of three early manuscript copies of Copernicus' *Commentariolus*. That called for an illustration, as did a curious paper instrument bound into one of its other copies. Next comes Edinburgh, with Adam Smith's copy, John Craig's book with his copies of Paul Wittich's marginalia, the original Jofrancus Offusius copy, and of course the greatest copy of them all, the one with Erasmus Reinhold's magnificent annotations, the description of which takes ten more *Census* pages.

At length I came to Glasgow, with its three first editions of *De revolutionibus*. I sought to illustrate a page with the writing of Willebrord Snell (1580–1626), the astronomer nowadays famous for Snell's law, the formula that expresses how light bends as it enters (or exits) from a glass lens. I hadn't made the final selection of illustrations when in 2001 I started composing the actual camera-ready copy for the *Census*, so I was in for one more surprise as I went down the home stretch. Folio 81 verso of the Snell copy has at the bottom of the page a nice sample of his hand, signed with the initials *Ru Sn*, indicating that the comment came originally from his father, Rudolph Snell. But as I looked more closely at the photograph, I realized that someone else, whom I had not identified, had written the notes in the left-hand margin.

UNLIKE MANY of the relatively unknown astronomers in this story—Reinhold, Wittich, Offusius—Gerardus Mercator has a comparatively high name recognition on account of his map projection. This is the rectangular grid of longitude lines that makes Greenland look as big as the

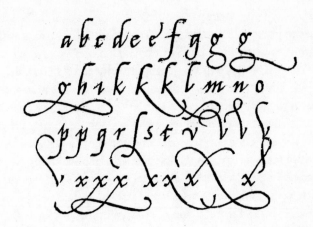

Âutumnali æquinoctio in Br
â Bruma in æquinoctiũ Vern
mæus, non aliter quàm ante fe
fe inueniffe teftatur. Quamob
pus, fummam abfidem xxiiii
& eccentroteta uigéfimamqua
quæ ex centro eft, perpetuo pe
nitur mutatum, differentia ma
ab æquinoctio Verno ad Æf
fcrup. xxxv. adnotauit: ad
clxxxii, fcrup. xxxvii.è qu
elicuit eccentroteta part. non a
tro eft 10000. Confentit huic
tis ratione, fed apogeũ prodid
x. quod Machometo Araïéfi

Gerardus Mercator's lowercase alphabet from his Litterarum *(Louvain, 1546) and a
sample of his italic hand in an annotation on folio 87 verso in his copy of*
De revolutionibus, *now in the Glasgow University Library.*

United States. Besides being a redoubtable cartographer, Mercator was a Renaissance polymath: an astronomer, astrolabe maker, engraver, and the man who revolutionized handwriting. Whenever you write a capital *E* in this shape: $\mathcal{E}$, you are using a form introduced by Mercator.

Mercator's is still a name to be reckoned with in Belgium, and in 1994 his fans (with the help of local banks) produced a lavish tome chronicling his accomplishments. I was delighted with the book, but what particularly caught my eye was an appendix that concerned his library. In the nineteenth century there still existed a single printed pamphlet listing his books, but this can no longer be found. Fortunately, however, someone had made a handwritten copy, which was typeset and printed in the splendid *Gérhard Mercator* volume. And there, among his mathematics books, is the entry for *De revolutionibus*, "cum annotationibus marginalis Gerardi Mercatoris."

By this time I knew of quite a few "missing" copies of Copernicus' book from references in early catalogs. There are, for instance, the two copies owned by the Elizabethan magus John Dee, who had one of the largest libraries in sixteenth-century England, though his library inventory does not reveal whether he annotated them or not. Similarly, the courtier Sir Walter Raleigh remarked in his *Historie of the World* (London, 1614) that he owned Copernicus' book, but again, there is no mention of whether he annotated it nor whether he wrote his name in his books. The same is true for the Astronomers Royal John Flamsteed and Edmond Halley.

But the Mercator record is different. It actually documents that he had written notes in his copy. Since the new Mercator volume included plenty of examples of his handwriting, I could systematically compare them with the microfilms and photographs that I had collected in which the handwritten notes were still unidentified. For example, a mystery copy is the well-annotated *De revolutionibus* owned by the famed eighteenth-century Scottish economist Adam Smith, who actually wrote a short essay on the history of astronomy. But the annotations aren't Adam Smith's. They predate him, firmly planted in the sixteenth century, and I wish I knew who wrote them because they are extensive and interesting. Alas, Mercator's hand didn't match. On several other mystery copies I had moments of high anticipation, only to realize in the cold light of reality that the Mercator samples did not provide a convincing attribution. So Mercator seemed destined to be another "missing person" as the *Census* headed toward its completion.

By what fates I know not, I decided to check the unidentified hand on folio 81 verso of the Snell copy sample against Mercator's hand, something I hadn't done previously because I mistakenly assumed that the annotations in the Snell copy were all accounted for. All at once I realized that I had at last pinned down the missing Mercator copy. A very characteristic and idiosyncratic tail of the letter *g* was absolutely convincing—those marginal notes clearly matched the cursive hand of the many missives illustrated in the big *Gérhard Mercator* book.

But that was not the end of the mystery. As I examined the group of photocopies from the Glasgow book, I realized that a third hand was involved. It wasn't Rudolph Snell's, and it wasn't Jean-Baptiste Colbert's, the French chancellor who owned the book after Snell. The annotations were clearly linked with Mercator's. I knew that the Louvain astronomer Gemma Frisius was Mercator's teacher, but the location of his copy was already known, in the Frisian library at Leeuwarden in northern Holland, and his handwriting was quite different. Johannes Stadius, a prolific calculator of Copernican-based ephemerides, was another member of the Louvain circle, but his copy, too, was accounted for—it ended up in the library of the West Point Military Academy. And the copy of yet another Netherlandish Copernican, Philips Lansbergen, is probably the one in Toronto, with quite different annotations.

Whose could the third hand be? Several days later, still feeling exuberant about the identification and recovery of the missing Mercator copy, I shared my discovery with a colleague, showing him both the photocopies and the lavish *Gérhard Mercator* volume. In describing what a polymath Mercator was, I turned to pages showing his astrolabes and then to those with a sample of his new italic letters. Suddenly I got an idea. What about comparing the italic letters with the third hand in the book? Almost immediately the match became clear. The capital *E*, *h*, and *p* were particularly distinctive. The third hand was also

Mercator's but written with his new letter forms. Even though in the end his notes themselves didn't seem particularly illuminating, identifying Mercator's copy brought a satisfying closure to one of the principal remaining mysteries of the Copernicus chase.

ONE MORNING as I was hard at work on the *Census*, I got a call from Jonathan Hill, a leading book dealer in New York City who had taken a keen interest in the project and who had given me many helpful pointers on the proper way to describe books as physical objects. "I've just obtained a nice copy of the second-edition Copernicus," he informed me.

"Well, have I seen it? Is it in the census?"

"How in the world would I know?" was his candid reply, so I suggested that he send the book up to Cambridge so I could inspect it, which he promptly did.

This was by no means an unprecedented request. I had once had a first edition in my office for nearly six weeks in late 1980, on loan for inspection from the San Francisco dealer Warren Howell, and I didn't even catch on that I had seen it before. Nearly two years after I had sent the copy back to California I got a call from Howell, who asked me to compare my notes on his copy with those from the copy I had earlier seen in the John Crerar Library in Chicago. Because all my census notes were in an extensive set of file folders in a row on my desk, I offered to handle his request in real time while he stayed on the line. I propped the receiver on my shoulder and dug into my notes. The Chicago copy, like Howell's, was censored in the standard way specified by Rome in 1620. But the Chicago copy had seventeen extra manuscript lines on folio 4 verso. Amazingly, so did Howell's copy. And both had exactly the same page size, to the millimeter.

"Why, these are the same book!' I blurted out with some surprise. "I didn't know the Crerar Copernicus was missing."

"That's what I was afraid of," Howell said rather sadly. "I didn't know either that it was missing. Don't tell anybody just yet. I've got to figure out what I'm going to do."

Howell, an eminent bookman whose specialties included rare science, had been approached by a man named Joseph Putnam claiming to be an émigré from Eastern Europe who had brought the books from there in a clandestine manner, and who had on offer some wonderful classics in the history of science, including *De revolutionibus* and the even rarer *De motu cordis* by William Harvey. Unknown to Howell, Putnam had ingratiated himself with the staff of the John Crerar Library in Chicago, so that he had regular unsupervised access to the rare book vault and had helped himself to very rare books. He went in with a large briefcase, sealed the books he wanted into large addressed envelopes as if they were finished projects ready to be mailed, and walked out of the library with only cursory inspections. While he was at it, he removed all the relevant cards from the card catalog as well. His clandestine operation began to unravel when someone inquired about a medical manuscript known to be uniquely in the Crerar Collection, but which Howell sold to another dealer and which ended up in the Staatsbibliothek in Berlin. In parallel the Crerar Library discovered its rare Harvey volume couldn't be found, and eventually they took an inventory and found their losses were much greater. But only when the FBI showed Howell a list of the missing books did he begin to appreciate that he was the victim of a major scam. That's when he called me to check if the Copernicus book really came from the Crerar. A few weeks later, in mid-January 1983, he called back to report that Joseph Putnam had been arrested in Milwaukee, and that he had more than 300 rare books from the Crerar Library in his house.

Howell tried his best to get the books back from his customers, with partial success, but he died a broken man in the following year—a tragedy, I felt, because he was such a kind and unmalicious person. Putnam was found guilty, and sentenced to two years in federal prison.

The second-edition *De revolutionibus* that Jonathan Hill had on consignment was a lot easier to recognize. Someone had taken a black felt-tip pen to delete one identifying word from the Latin inscription on the title page, but a computer search quickly picked up the rest of the phrase. The marked-out word was *Brunae*, Latin for Brno. I had inspected the book in the State Scientific Library in Brno, but after communism faded and Czechoslovakia fissioned, the library had returned the book to its former owner, the Augustinian Monastery outside Brno. It was a well-known establishment, where the renowned geneticist Gregor Mendel had been the abbot.

Now the question was: Had the monastery deaccessioned the book to gain some needed cash? Hill had the book on approval from a reputable German dealer, who was in turn convinced that his source was legitimate. The Web and e-mail having come into ascendancy, I could confirm the status of the book within thirty-six hours, and the news was bad. The monastery had not sold the book. It had disappeared along with a temporary employee. I assume the book is now back in the Brno monastery, and that someone lost money to the vanished thief. Whether it was the German dealer or his insurance company I do not know, but I am aware that nowadays many dealers carry insurance policies for precisely this sort of situation.

Unfortunately, it seems that in the majority of cases, unlike these two, stolen books do not find their way home.

THE COMPUTER typesetting for the 112 first and second editions now in the United States was relatively easy, since with few exceptions the descriptions and annotations were quite straightforward, even though the block of American holdings was the largest single group. By and large the American copies have neither the extensive annotations nor the extended provenances that some of the European copies boast; precisely why this is true is hard to divine. The most conspicuous exception is the Yale Beinecke Library copy, whose description goes for eight pages including five illustrations. No other American description exceeds an entire page except for a first edition at the Linda Hall Library in Kansas City, whose brief description includes a full-page plate to illustrate a small paper instrument laid in the book. There are, of course, besides Yale's "three-star" copy, a few "two-star" exemplars, including the first-edition Offusius copies at the University of Oklahoma and the University of Michigan.

Eventually, I reached the final U.S.A. entry, a second edition at the University of Virginia, perhaps the copy ordered for the library by

Thomas Jefferson. Because many of the volumes of the library were destroyed by fire in 1895, it is now impossible to know whether the copy is the original one or a replacement.

Since Yugoslavia had been replaced by Croatia, it had lost its final place in the alphabetic rankings, and since Vatican City had been grouped with Italy, the sequence closed with "USA: Texas–Virginia." This left just the appendixes, an assortment of miscellaneous tables such as a chronological list of auction records (which nicely documented the ever-upward price of the two editions), and lists of locations of the related Copernican books, namely, Rheticus' *Narratio prima* and his 1542 edition of the trigonometric section of *De revolutionibus*. Though I had long been collecting locations of this latter book, I belatedly noticed that I had never systematically organized a list. Disconcerting as this last-minute realization was, I decided that I would include such an appendix if within two days I could get up a list of approximately forty libraries holding copies of the *De lateribus et angulis triangulorum*.

I had located over two dozen copies of this work as I visited libraries in search of *De revolutionibus* itself, and I got almost a dozen from the NUC (National Union Catalog of the Library of Congress) and the OCLC (On-line Computer Library Center). The richest haul came from the Karlsruhe Virtual Catalog, the main European computer database, enough to put me well over my goal. Another source was the Italian bibliographer Giovanna Grassi's published list of early books in European astronomical observatory libraries. Glancing at her listings for Copernicus, my eyes wandered from *De lateribus* to *De revolutionibus*, and I was aghast! Though her book had been on my shelf for more than a decade, and was a reference I had frequently consulted, for some reason I had never checked it for locations of the magnum opus itself. Grassi listed fifteen copies of the first edition, fourteen of which I had seen, but one that was new to me. Similarly, of the fifteen copies of the second edition, there was also a new location. It appeared that the observatories in Naples and in Athens had copies of *De revolutionibus* I had overlooked.

With long-distance telephone and e-mail, the confirmations were in hand almost instantly. But there was still more news. The Naples observatory had not only the first edition but the second as well. The three additional copies, now recorded in an addendum to the *Census*, brought the total number of entries to 601. That total included three German copies destroyed in World War II, for which some limited information about their early provenances has survived. There were as well a few far-flung copies that I did not personally inspect, in Lithuania, Corsica, Croatia, and the Philippines, for example, although I received information about all these copies. And the Vatican librarians have not been able to find one of their copies, though for completeness I listed it without any details.

How many extant copies did I fail to list? An unanswerable question, of course. As the survey had gone on, the discovery of new locations had dwindled to a small trickle, yet if I extrapolated, the number always went to infinity because there was no evidence that I had come to the end of previously unknown copies. Infinity has to be the wrong answer, of course, since the likely number of first-edition copies printed lies between four and five hundred, and for the second edition between five and six hundred. Since the *Census* was printed, I have located one more copy of the first edition and six more of the second. If I had to pick numbers, I would estimate that there may be as many as a dozen undiscovered copies of the first edition and two dozen of the second. Only time will tell how cloudy my crystal ball really is.

From the earliest days of my Copernican census I had conferred with my Polish colleagues and had agreed with Paul Czartoryski, editor in chief of the *Studia Copernicana* series of the Polish Academy of Sciences, that the *Census* should be published in their series. As the project grew in scope, I realized that the volume would be of interest to librarians and book collectors all over the world, and that it would be highly desirable to copublish with a Western press to guarantee its ready availability. Eventually, Czartoryski organized an agreement with Brill Academic Publishers in Leiden for a Western *Studia Copernicana* series, and thus they became the publisher. Brill did a splendid job, printing the book on high-quality paper that works well with the sixty-three glossy images it contains.

Brill published the *Census* in an edition of only four hundred, just matching the lower estimate of the first edition of *De revolutionibus*, and guessed that it would take twenty years to sell that many copies, again approximately matching the time it took Petreius' edition of the Copernicus to sell out. These estimates missed rather badly, because the volume is already almost out of print.

Only a handful of books have been targets for a complete census. Of course bibliophiles are keen to keep a census of all the locations of Gutenberg Bibles. As recently as 1996 a previously unlisted fragment of 177 pages was found in Rendsburg, Germany, but that seems unlikely to be repeated. It was comparatively easy to compile a census of copies of John James Audubon's *Birds of America*—not just because the giant elephant folios are hard to miss but because there was a list of original subscribers. A majority of all known copies of the so-called First Folio of Shakespeare's plays resides in the Folger Shakespeare Library in Washington, and a century ago a census of this title was made. None of these efforts includes the sort of detailed physical descriptions and lists of provenances found in the Copernican *Census*, although such a project is currently under way for the Shakespeare First Folios.

An extensive list has been made of locations of Isaac Newton's *Principia*, without attempting to be definitive. That highly significant volume would be a good target for a survey comparable to the Copernican census, but as my own experience has shown, such a labor cannot be undertaken lightly. An excellent listing of locations of all of Johannes Kepler's titles recently appeared, but with no attempt to locate copies in private collections. His *Astronomia nova* of 1609 would be another candidate for a survey of annotations, but my own preliminary search suggests it would not be nearly as interesting as the Copernicus chase because there seem to be many fewer annotated copies.

ONE MIGHT examine the series of provenances that the *Census* lists for each book in order to find which copy was once owned by a movie star, which by a saint, or which by a heretic. Or which copy has the longest list of provenances. Many copies have only a single provenance simply be-

Title page of De lateribus et angulis triangulorum, *the trigonometric section of Copernicus' book that was edited by Rheticus in Wittenberg in 1542.*

cause the previous owner is completely unknown. In a few cases, of course, the copy is still in the original institution that acquired it in the sixteenth century. But some copies have records of remarkable journeys.

I am amazed at the travels of one of the first editions now at the Paris Observatory. The earliest evidence for its ownership dates from the second half of the sixteenth century when an unidentified owner copied out the notes from Jean Pierre de Mesmes' copy; De Mesmes had in turn during the late 1550s copied most of his notes from his teacher, Jofrancus Offusius. That anonymous owner included some of his own notes, including a citation to Maestlin's *Ephemerides*, which were published in 1580. The next record comes from Caspar Peucer, who had been a student of Erasmus Reinhold in Wittenberg and who succeeded him as professor of astronomy there in 1554. Despite his insider's position—he had married Philipp Melanchthon's daughter in 1550—he was jailed for his crypto-Calvinism for a dozen years between 1574 and 1586, during which time he made ink from his own blood in order to continue writing. He was subsequently rehabilitated and played host to Tycho Brahe when the Danish astronomer was on his way from Denmark to Prague. Peucer must have got this *De revolutionibus* after he was released from jail, though how it got from Paris to Wittenberg remains a blank spot in its history. The next known owner of this copy was Joseph-Nicolas Delisle, a French astronomer who founded the original observatory in St. Petersburg in the eighteenth century and who amassed a substantial personal library, which went to the library of the Depôt de la Marine and subsequently to the Paris Observatory, where it has been ever since. We can imagine that Delisle found the book in Germany, took it along to Russia, and eventually brought it back to Paris, but that is of course pure speculation. Precisely where and when he acquired the book is again an unknown. In any event, I still find it astonishing that the book returned to France after such a journey.

A more fragmentary provenance, but nevertheless one that I was particularly smug about establishing, links the first edition now in the Scheide Collection at Princeton with the large Biblioteca Nazionale Braidense in Milan. Three generations of Scheides have formed this private collection,

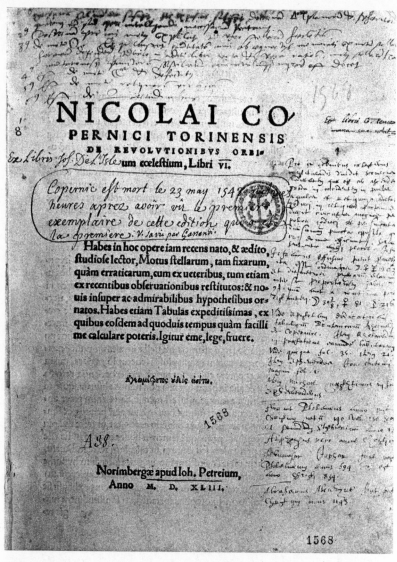

Perhaps the most heavily annotated title page known, in a Paris Observatory first edition of Copernicus' book, including an attribution of notes to Caspar Peucer, Reinhold's successor at Wittenberg (upper right), and the hand of Joseph-Nicolas Delisle, who brought the book back to Paris (left center).

which is housed in the Princeton University Library. John Hinsdale Scheide, the founder of the collection, bought his copy in 1928 for $785 from the colorful, pace-setting American dealer A. S. W. Rosenbach. It contained some illegible initials, the manuscript date 12 March 1710, and an old but unrelated shelf mark, S.15.11, but otherwise no identifiable provenance. When I visited the National Library in Milan, not only did I record its two second editions—one censored according to the Inquisition's decree, the other totally unannotated—but I also noticed in the old manuscript catalog of the library that at one time this institution had a first edition with the shelf mark S.XV.II. One day in my office, much later, when I was idly sifting through some of the census data, it occurred to me to try to compare this shelf mark against some of the old numbers I had recorded on books in other collections, and almost immediately I found the match. Surely, the Scheide copy was once in the Milan library, but how and when it got out will probably forever remain a deep mystery. One can only wonder if Rosenbach, a truly remarkable wheeler-dealer, had any clue about its origins.

I'm always curious about published catalogs of old collections, and when I get the chance, I examine them to see if they include Copernicus' book. By and by I encountered *Biblioteca Firmiana*, a multivolume library catalog published in Milan in 1783. The collection, established by Carlo Giuseppe Firmian, included a first-edition *De revolutionibus*, presumably the one that subsequently entered the Braidense collection. That chance encounter with the Firmiana catalog established a still earlier provenance for the Scheide copy, but neither the Scheide Copernicus nor the Paris Observatory copy come close to holding a record number of provenance steps. The winners in the *Census* each have nine listed ownerships: a second edition in the Ossolineum in Wrocław, and a first edition at the University of Michigan in Ann Arbor.

Frankly, the provenances of the Polish copy are not very enlightening, because mostly they are unidentified or even almost illegibly blotted or trimmed off the page. The Ann Arbor copy tells a better story, but also there the earliest steps are obscure, though the original (and unidentified) annotator got the volume off to an illustrious start in the late 1550s by transcribing the marginalia from Offusius' copy (the one now in Edinburgh). The book later found its way into the distinguished Lamoignon family collection in Paris and appears in their printed catalog of 1770—the catalog itself being a considerable rarity since it was printed in an edition of only fifteen copies! Eventually, the collection was obtained by an English bookseller, Thomas Payne, who auctioned the *De revolutionibus* in 1791. Its location was unrecorded for the eighteen years from 1791 until 1809, when it was acquired by Stephen Peter Rigaud, Savilian Professor of Astronomy at Oxford. After his death Rigaud's collection was purchased *en bloc* by Oxford's Radcliffe Observatory, but in 1935 the observatory's library was turned over to the Bodleian Library, which auctioned the duplicates. Through an agent, Tracy William MacGregor, a Detroit philanthropist, bought the book for £165, and three years later he presented it to the University of Michigan Library.

Four other copies show a sequence of eight ownerships, two of which begin in the sixteenth century. One is a first edition at Trinity College, Cambridge, which I described in chapter 2. The other, a second edition with a very speckled career, has come to rest, quite appropriately, at the university library in Wrocław. This is one of several copies annotated by Paul Wittich, but not one of the three that were later purchased by Tycho Brahe, because it apparently stayed around Wrocław (Wittich's hometown) long enough for a detailed copy of the annotations to be made by a young fellow townsman, Valentin von Sebisch. The Wittich copy was for a while in the library of the Pollinger College (probably in Bavaria), according to an inscription that I deciphered using my ultraviolet light. A considerably defaced inscription indicates that in 1816 the book was owned by the Russian Baron von Canstadt, who spent much time in Munich; at some point the book went into the Royal Library there, which eventually disposed of the copy as a duplicate. The copy was acquired and rebound by the Royal Astronomical Society in London. Since the copy lacked Rheticus' *Narratio prima*, which had been included in the published second edition but presumably removed at the Pollinger College because Rheticus was a prohibited author, the Royal Astronomical Society released it as a defective duplicate in 1949, not realizing that it contained one of the most important extant sets of sixteenth-century annotations. The copy was held for some years by the dealer Ernst Weil before Roman Umiastowski eventually bought it. In due time I persuaded Umiastowski that it would make a fine gift to the university library in Wrocław, and he presented it to the library in 1982. This gift was doubly appropriate, not only because Wittich came from Wrocław but because Sebisch's copy is also in the university library there.

The census turned up a splendid parade of owners. Copies were owned by the Escorial architect Juan de Herrera, the astronomer-cartographer Gerardus Mercator, the Venetian music theoretician Giuseppe Zarlino, the Pléiade poet Pontus de Tyard, the humanists Johannes Sambucus and Petro Francesco Giambullari, the antiquaries John Aubrey and William Camden, and the financier Johann Jacob Fugger. Henry II of France, Philip II of Spain, George II of England, Sigismund II Augustus of Poland, Count Egmont, and Elector Otto Heinrich had the book in their collections, as did Duke August, whose library at Wolfenbüttel was the finest in Europe in the early eighteenth century. Among the early owners were Saint Aloysius Gonzaga, Giordano Bruno, John Dee, Thomas Digges, Tycho Brahe, Galileo, Kepler, and a host of lesser-known medical doctors, astrologers, and dilettantes, including more recently the Hollywood actor Jean Hersholt, whose second edition is now in the Library of Congress.

Of course, not all of these owners actually read the book. The royalty did not annotate their copies, but many others did, leaving behind a precious legacy of the way in which the book was perceived and read during the scientific Renaissance. Clearly, when Arthur Koestler wrote that *De revolutionibus* was "the book that nobody read" and "an all time worst seller," he couldn't have been more mistaken. He was wrong.

Dead wrong.

EPILOGUE

"WHAT'S GOING on here?" The inquisitor was Don Goldsmith, a confidant and sometime intellectual sparring partner. He had unexpectedly arrived from Berkeley one afternoon in 2000 just as two previous visitors were leaving my office.

"Oh," I said, playing dumb, "why do you think anything special's going on?"

"Well, your ordinary academic visitors don't have revolvers bulging in their pockets."

Discretion was the order of the day, so I could only hint at the circumstances. He had surmised correctly that the FBI had paid a call, and he undoubtedly suspected it had something to do with rare books and probably with Copernicus, but I couldn't reveal the details then.

A few months earlier Christie's in London had announced that a first edition of *De revolutionibus* would be auctioned in their rooms, with an estimated bid of $500,000–$800,000. Scarcely a year before the auction date, a "reader" had ordered up a stack of rare books at the university library in Kiev, and at some point had gone out for a smoke, leaving his jacket and the stack behind. But when closing time came, he hadn't returned, and the stack of books was one volume short, lacking the 1543 *De revolutionibus*. Six months later a copycat theft took place at the Academy of Sciences Library in Cracow. Suddenly Christie's found itself under siege. The Polish press blossomed with stories that their Cracow Copernicus was about to be auctioned in London, and the Polish authorities put in a claim for it. I assured the Christie's experts that their copy was entirely new to me and therefore not one of the recently stolen books, and I could testify that it did not match the missing Cracow copy, whose description I had carefully recorded. They relayed the message to Poland, but to no avail. Two Polish librarians flew to London to see for themselves that the book was not theirs. Their trip was not in vain, however, because they recovered a stolen copy of Galileo's *Sidereus nuncius* that was also scheduled for the auction.

With just two days to go before the sale, Scotland Yard reported that another claim was being staked, this time by Albania. This threw me into some panic, as it had never occurred to me that any library in Albania would have a copy, especially because, even though it had been a tightly closed country, it was known to be impoverished and with relatively few intellectual treasures. Albania turned out to be a red herring, because the message had gotten garbled. It was the Ukraine making a claim for the missing Kiev copy. I assured the Christie's experts that I had good records on the Kiev copy, and that it was not the copy they were intending to sell. Then, less than twenty-four hours before the auction, a letter arrived from lawyers at the Massachusetts Institute of Technology stating that they thought the book might be theirs. Believing that the sale had been poisoned, Christie's withdrew the book.

In a rather bizarre way my researches had paved the way for what the MIT librarians naively thought was a reasonable claim. An eighteenth-century owner of the Christie's copy had been one Louis Godin, a French astronomer who had gone to Peru to measure the arc of the meridian,* and who later became the director of a Spanish academy in Cádiz. I had told the Christie's experts that a second-edition *De revolutionibus* also owned by Godin was currently at the Dibner Institute at MIT. Subsequently, the Christie's auction description for the first edition included the fact that another copy once owned by Godin was at MIT, but it didn't mention the specific location.

Sharp-eyed MIT librarians checked their catalog and failed to find

---

*A way to determine the detailed shape of the Earth's globe.

such a book because the Dibner Institute is an independent organization on the MIT campus, and its considerable collection of rare science books isn't listed in the main MIT library catalog. But with a little more checking the librarians turned up a disconcerting fact: In 1924 MIT had been given a first edition, which was simply placed on the open shelves; by 1932 it could no longer be found. Like a slumbering giant awakened, the institute had sleepily lashed out, threatening a hold on the Christie's copy. Fully awake, and after the librarians had an opportunity to browse through my records to see how many first editions had been auctioned between 1932 and 2000, they realized the claim was ill-conceived, and eight months later Christie's at last successfully auctioned its first edition with the Godin provenance. It fetched $500,000.

At this point several international auction houses were asked by a Russian agent if they wished to auction another copy. The staff at Christie's, by now considerably educated about the perils of selling *De revolutionibus*, replied that they would consider it only if I had checked it out in advance. Very soon I received, via e-mail, digital images of several pages from another first edition. The amateur pictures were distressingly fuzzy, but I could just barely decipher some names of early owners written on the title page. For years I had retained a fading mental picture of every *De revolutionibus* I had seen, but eventually, of course, the entire database was transferred to a computer, which made searches a great deal faster and more reliable. Checking my files, I quickly ascertained that I had previously examined the copy in the Russian Academy of Sciences Library in Leningrad. This was the point at which the well-armed FBI agents visited my office.

As it turned out, there was no evidence that the stolen copy had left Russia, so there the matter rested. Presumably, the appropriate Russian authorities were notified, but nothing happened. After a couple of months passed, I contacted an astronomer in St. Petersburg and asked him to inquire after the academy's copy. Within a week the international press carried the news: Twenty rare books were missing from the vaults of the academy library. Apparently, the director had been unaware of the losses, which certainly smacked of an inside job. Curiously, but not without precedence, no public announcement indicated which books had been stolen. Soon, however, the FBI brought me the list. I was shocked to discover that not one but two first editions of *De revolutionibus* had been part of the haul.

Four copies had gone missing in two years. In addition, there were the earlier thefts of first editions I had seen at the Mittag-Leffler Institute in Stockholm, and at the University of Illinois Library in Urbana-Champaign. Sad to say, none of these copies has been recovered, leaving me with the dubious distinction of having personally inspected more copies of the first edition than can now be located.

But a happier adventure spun itself out over a quarter of a century of investigating copies of *De revolutionibus*. In 1974 I had undertaken a field expedition to a series of provincial libraries in Italy, including the Biblioteca Palatina in Parma. There I saw a second edition of *De revolutionibus*, but unfortunately its first edition was nowhere to be found. The librarian even showed me the gap on the shelf where it was supposed to be. I assumed that the volume was simply in use in some library office and gave it no further thought. Sometime later when I happened to tell Robert Westman about the missing book, he remarked that, curiously enough, it had been missing when he tried to see it a year earlier than I had. This information aroused my interest in the book, so I wrote to the library asking for the description of the volume as it appeared in their catalog. My vague memory was that the catalog had mistakenly attributed the Andreas Osiander introduction to the great sixteenth-century classicist Joachim Camerarius. When the description arrived, I realized that I had hastily misread it when I had visited Parma; what the citation actually indicated was that the volume was prefaced by a Greek manuscript poem in the hand of Camerarius.

I knew of one book, undoubtedly the most fabulous copy in private hands, that matched this description. It had turned up, mysteriously, in the London book market shortly after World War II. Could that copy have been "liberated" by an Allied soldier or by a hungry librarian? Because its then current owner, the extraordinary collector Haven O'More, lived part of the year in Cambridge, Massachusetts, I asked him if he would bring his precious book into my office at the observatory. When he did, one afternoon in March 1981, we took it to a dark closet and inspected it with ultraviolet light. The rag paper used in sixteenth-century editions glows under the UV except where ink has been applied. Even if the ink has been carefully washed out, the fluorescence is destroyed, and the writing will still show clearly in contrast to the glowing paper. Similarly, the glue marks from removed labels reveal themselves in telltale traces. But there were none. The book with the Camerarius manuscript poem appeared to have a clean bill of health—no identifying ownership marks had been erased or washed out. Nonetheless, because the description in the Palatina catalog seemed a perfect match, I was prepared to link the book brought to my office that afternoon with the missing Italian copy.

An unexpected episode a few years later taught me to be more cautious in making such inferences. Pierre Berès, an eminent French book dealer, informed me that he had something of interest concerning Copernicus. When I had occasion to go to Paris, I arranged a visit. Berès then related the following curious tale. An anonymous caller, possibly Italian, had inquired by phone if Berès was interested in some old books, including a first-edition *De revolutionibus*. When an affirmative reply was forthcoming, the mystery voice said, "You can't contact us, but we'll get in touch with you." Some days later, Berès explained, a packet of photocopies arrived without a return address. The sheets were mostly copies of title pages, but with that of Copernicus' book were two xerographic copies recording a Greek manuscript poem, and these he produced for me. I saw at once that here was another copy of Camerarius' poem, so I explained the situation at the Biblioteca Palatina.

"This is surely the stolen Palatina copy," Berès quickly concluded, and he declared that he would abandon the pursuit of the book. With that the trail went cold, leaving me with a troubling dilemma: whether to publish the photocopies in my *Census* to alert possible buyers to a stolen copy, or to suppress this information lest the thief simply destroy the identifying sheet with the Greek poem.

After a decade passed, and before I had to make a final decision, the book prefaced with Camerarius' manuscript surfaced once again. This time a New York dealer had caught wind of its existence and had promptly alerted me. Again I said, "On guard!"

A few weeks later I was telephoned by an expert at Sotheby's in Milan. The Copernicus volume, unbound and in a somewhat disreputable state, had been brought to the auction house by a family in Parma. The book had been acquired by the father, since deceased, who had had a reputation as a local bibliophile, and who apparently did not raise questions about the sources of the books offered to him. What evidence did I have, the voice from Milan inquired, that the book belonged to the Palatina?

I explained about the catalog description. Apparently I had had access to a special catalog, not the public one, for my information was news to the Sotheby's representative. He nevertheless reported that he had investigated and discovered that most, if not all, of the books being offered to Sotheby's were listed in the Palatina catalog but were in fact missing from that library, strong circumstantial evidence for their true ownership. However, the absence of specific physical evidence created a sticky problem.

There the matter rested for several months. Then, in January of 2000, I received an e-mail stating simply, "Nicolaus is back in the Biblioteca Palatina." I'm not sure precisely how this happened, but it was a fitting conclusion to the dogged, adventurous, almost quixotic thirty-year pursuit into the way sixteenth-century astronomers reacted to the most revolutionary scientific advance in more than a millennium.

From Equant to Epicyclet

CELESTIAL MOVEMENTS are everlasting, Copernicus declared in the title of the fourth chapter of his *De revolutionibus*. And only a circle has neither beginning nor ending and can repeat what is past. Although the observed motions are complex, "it is impossible that a heavenly body should be moved irregularly by a single sphere." This could only be accomplished by an inconstancy of the moving power that drives it, and "the mind shudders" at such a prospect. Although Copernicus did not here single out Ptolemy's equant as an offense, clearly this is what he had in mind.

Ptolemy's equant produced uniform motion about an imaginary point within the circle that carried the planetary epicycle, causing the epicycle to move around that circle faster on one side than the other. Today we recognize the equant as an ingenious approximation to one of nature's most fundamental laws, the conservation of angular momentum. Without some device to account for its observed effects, the predicted positions of Mars, for example, would be wrong by many degrees.

For Aristotle, a hard-core geocentrist, with the Earth solidly fixed in the center, the motions originated at the outer edge of the system. The love of God turned the spheres, and the moving power was transmitted more and more slowly from one sphere to another, so that the inner sphere carrying the Moon took nearly twenty-five hours to turn once, whereas the stars spun about more quickly in twenty-four sidereal hours. Kepler, a confirmed heliocentrist, saw the implications more clearly than his Polish master: Mercury, the innermost planet, sped around the Sun the fastest, and therefore the driving moving power had to come from the inside out, somehow residing in the Sun itself. Because the Earth was in winter closest to the Sun, it

263

should soak up that moving power more efficiently then and should move faster. Kepler's mind did not shudder at such inconstancy; he thought it was physically logical.

Copernicus stood at a crucial transition point as he revised the geometric blueprint but was unable to come to terms with the physical implications of the radical realignment. Within his own aesthetic vision he had to generate the observed unequal motion out of combinations of uniform circular motions. Because this part of his cosmology turned out to be a physical dead end, modern secondary sources tend to ignore this rather complicated part of his *De revolutionibus*, even though it comprised the bulk of the book. Nevertheless, it proved to be the most intriguing and most studied part of his work as far as sixteenth-century astronomers were concerned.

Somehow, somewhere, Copernicus stumbled onto the way to generate Ptolemy's equant motion by using either a concentric circle with two small epicyclets, or an eccentric circle with a single epicyclet. To introduce it this way is to conceal the iceberg below its tip. In the mid-1950s the historian of Islamic science, Edward S. Kennedy, and his students at the American University of Beirut (where I was also then teaching) showed that a series of thirteenth- and fourteenth-century Muslim astronomers in Persia and in Damascus had come up with exactly the same arrangement of epicyclets that Copernicus used, although they did not proceed to a heliocentric cosmos. In fact the two aesthetic ideas are not coupled except in some inscrutable way in the mind of Copernicus. Ever since Kennedy's work a big question has been: Did Copernicus invent the epicyclet arrangement independently, and if not, how in the world did he find out about it? (There was no available printed source, and apparently no manuscript source in Latin.) One interesting guess is that Copernicus found out through the work of a sometime ephemeris maker named Johannes Engel, who quite likely used the device, and that Copernicus himself was probably unaware of its Islamic antecedents.

The double epicyclet with the concentric circle was Copernicus' system of choice when he wrote his *Commentariolus* and it is the "second method" of *De revolutionibus*, whereas the single epicyclet was the "third method" of *De revolutionibus* and was the one actually employed there. It really doesn't matter, Copernicus implied, because both as well as the first arrangement gave the same result, and therefore one of them had to be real—a consummate demonstration of faulty logic! In any event, to use the twentieth-century proverb "A picture is worth a thousand words," here is a diagram to show the equivalence of Copernicus' single epicyclet and Ptolemy's equant.

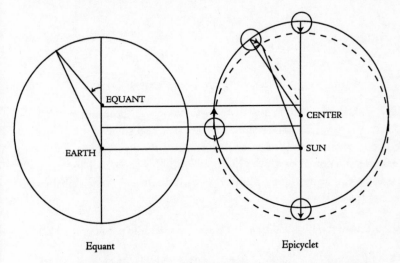

Equant   Epicyclet

Copernicus' replacement of the equant by a pair of uniform circular motions.
The epicycle always moves to form an isosceles trapezoid.

Note that the motion in the epicyclet is reckoned with respect to the moving line from the center of the circle, and the angle at the center matches the angle in the epicyclet. Because the radius of the epicyclet is one-third of the distance from the Sun to the center, a regular trapezoid results, and the dashed line replicates the uniform motion around the equant.

The path that results, shown by the dashed circle, is in fact not quite a circle because it bows out slightly on the sides—the opposite way compared to Kepler's ellipse.

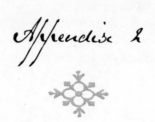

Appendix 2

LOCATIONS OF *DE REVOLUTIONIBUS*

THE TWO LISTS that follow do not exactly match those in *An Annotated Census of Copernicus' De Revolutionibus* because here some of the private collections have been omitted as have copies that have been sold to undisclosed locations and copies destroyed during World War II. In each list the copies are listed alphabetically by country, and within the countries alphabetically by city (or by state).

Locations of the First Edition of *De revolutionibus* (Nuremberg, 1543)

AUSTRIA
Vienna, Österreichische
Nationalbibliothek
Vienna, Universitätsbibliothek
Vienna, Universitätssternwarte

BELGIUM
Brussels, Bibliothèque Royale
Albert 1ᵉ (2 copies)
Brussels, Observatoire Royale
Liège, Bibliothèque de
l'Université

CANADA
Montreal, Quebec, McGill
University, Osler Library

CZECH REPUBLIC
Olomouc, Státní Vĕdecká
Knihovna
Prague, Knihovna Národního
Muzea (National Museum
Library)
Prague, Strahovská Knihovna

DENMARK
Aarhus, Statsbiblioteket
Copenhagen,
Universitetsbiblioteket
(2 copies)

FRANCE
Besançon, Bibliothèque
Municipale

Bordeaux, Bibliothèque
  Municipale
Brest, Bibliothèque de la Marine
Dijon, Bibliothèque Municipale
Évreux, Bibliothèque
  Municipale
Grenoble, Bibliothèque
  Municipale
La Rochelle, Bibliothèque
  Municipale
Lille, Bibliothèque Municipale
Lyons, Bibliothèque Municipale
  (2 copies)
Marseille, Bibliothèque
  Municipale
Metz, Bibliothèque Municipale
Paris, Bibliothèque de l'Arsenal
  (2 copies)
Paris, Bibliothèque de l'Institut
Paris, Bibliothèque de
  l'Observatoire (2 copies)
Paris, Bibliothèque Mazarine
Paris, Bibliothèque Nationale
  (3 copies)
Paris, Biblioteka Polska
Paris, Société Astronomique de
  France
Poitiers, Bibliothèque
  Universitaire de Poitiers
Soissons, Bibliothèque
  Municipale
Strasbourg, Bibliothèque
  Nationale et Universitaire
Toulouse, Bibliothèque
  Municipale

Troyes, Bibliothèque
  Municipale
Vienne, Bibliothèque
  Municipale

GERMANY
Aschaffenburg, Stiftsarchiv-
  Bibliothek
Bonn, Universitätsbibliothek
Braunschweig, Stadtbibliothek
Darmstadt, Hessische Landes-
  bibliothek (2 copies)
Dillingen a, D. Donau,
  Studienbibliothek
Dresden, Sächsische
  Landesbibliothek (2 copies)
Erlangen, Universitätsbibliothek
Freiburg im Breisgau,
  Universitätsbibliothek
Giessen, Universitätsbibliothek
Gotha, Forschungsbibliothek
  (2 copies)
Göttingen,
  Universitätsbibliothek
Heidelberg,
  Universitätsbibliothek
Jena, Universitätsbibliothek
  (2 copies)
Karlsruhe, Badische
  Landesbibliothek (2 copies)
Kiel, Universitätsbibliothek
Leipzig, Universitätsbibliothek
Lindau, Stadtbibliothek
  (2 copies)
Mainz, Stadtbibliothek

Munich, Bayerische
Staatsbibliothek (2 copies)
Münster, Universitätsbibliothek
Neresheim, Bibliothek der
Abtei
Nuremberg, Landeskirchliches
Archiv
Passau, Staatliche Bibliothek
Pommersfelden,
Schönbornsche Bibliothek
Potsdam-Babelsberg,
Sternwarte
Rostock, Universitätsbibliothek
Schulpforte bei Naumburg,
Oberschule Pforte
Schweinfurt, Stadtbibliothek
Stuttgart, Robert Bosch
Stiftung
Stuttgart, Württembergische
Landesbibliothek
Tübingen,
Universitätsbibliothek
Ulm, Stadtbibliothek
Weimar, Herzogin Anna
Amalia Bibliothek
Wolfenbüttel, Herzog August
Bibliothek (2 copies)
Würzburg,
Universitätsbibliothek
Zittau, Christian-Weise-
Bibliothek
Zwickau, Ratsschulbibliothek

HUNGARY
Budapest, Egyetemi Könyvtár
(University Library)

Debrecen, Library of the
Transtibiscan Church
District

IRELAND
Dublin, Trinity College

ITALY
Bologna, Biblioteca
Universitaria
Florence, Biblioteca Nazionale
Centrale
Naples, Biblioteca
Universitaria
Naples, Osservatorio
Astronomico di
Capodimonte
Padua, Biblioteca Universitaria
Palermo, Biblioteca
Nazionale
Parma, Biblioteca Palatina
Pisa, Biblioteca Universitaria
Ravenna, Biblioteca Classense
Rome, Accademia Nazionale
dei Lincei
Rome, Biblioteca Nazionale
Centrale (2 copies)
Rome, Osservatorio Romano
Turin, Biblioteca Nazionale
Universitaria (3 copies)
Venice, Biblioteca Nazionale
Marciana
Vicenza, Giancarlo Beltrame
private collection

JAPAN
Fukushima (Iwaki City), Iwaki
Meisei University
Hiroshima, Hiroshima
University of Economics
Kanazawa, Kanazawa Institute
of Technology
Kyoto, Kyoto Sangyo
University
Osaka, Kinki University
Tokyo (Hino City), Meisei
University

LITHUANIA
Vilnius, Vilnius University
Library

MEXICO
Guadalajara, Biblioteca
Publica

NETHERLANDS
Amsterdam, Bibliotheca
Philosophia Hermtica
Amsterdam,
Universiteitsbibliotheek
Leeuwarden, Provinciale
Bibliotheek van Friesland
Leiden, Universiteitsbibliotheek
Zutphen, St. Walburgskerk
Librarij

PHILIPPINES
Manila, Santo Tomás
Universidad

POLAND
Cracow, Biblioteka Jagiellońska
(2 copies)
Cracow, Biblioteka Polskiej
Akademii Nauk (copy stolen)
Cracow, Muzeum Narodowe
(Czartoryski Library)
Kórnik, Biblioteka Kórnicka
Płock, Towarzystwo Naukowe
Płockie (Płock Scientific
Society)
Poznan, Biblioteka Raczńskich
Poznan, Poznanskie
Towarzystwo Przyaciół Nauk
Toruń, Biblioteka
Uniwersytecka
Toruń, Książnica Miejska im.
Kopernika
Warsaw, Biblioteka Narodowa
Wrocław, Biblioteka
Uniwersytecka (2 copies)
Wrocław, Ossolineum

PORTUGAL
Lisbon, Academia das Ciências
Lisbon, Biblioteca da Ajuda
Porto, Biblioteca Pública
Municipal

RUSSIA
Moscow, Russian State Library
(3 copies)
Moscow, University Library
St. Petersburg, Academy of
Sciences Library
(2 copies stolen)

St. Petersburg, National
Library of Russia
St. Petersburg, Pulkovo
Observatory

SPAIN

El Escorial, Biblioteca de San
Lorenzo
Madrid, Biblioteca Nacional
Madrid, Placido Arango
private collection
Madrid, Palacio Real
San Fernando, Observatorio de
Marina
Seville, Biblioteca de
Universidad
Valencia, Biblioteca de
Universidad

SWEDEN

Djursholm, Institut Mittag-
Leffler (copy stolen)
Stockholm, Ingenjösvetenskap-
sakademiens
Uppsala, Universitets Bibliotek
(2 copies)
Västerås, Stadsbibliotek,
Diocese Library Collection

SWITZERLAND

Basel, Universtätsbibliothek
Bern, Stadt- und
Universtätsbibliothek
Geneva, Bibliotheca Bodmeriana
St. Gallen, Kantonsbibliothek
Vadiana

Schaffhausen, Stadtbibliothek
Zurich, Eidgenössische
Technische Hochschule-
Bibliothek

UKRAINE

Kiev, National Library of the
Ukraine (copy stolen)
L'viv, University Library

UNITED KINGDOM (ENGLAND)

Cambridge, Christ's College
Cambridge, King's College
Cambridge, Peterhouse
Cambridge, Trinity College
(3 copies)
Cambridge, University Library
Chatsworth, Duke of
Devonshire Collection
Deene Park, E. Brundenell
Collection
Eton, Eton College (2 copies)
Greenwich, National Maritime
Museum
Liverpool, University of
Liverpool (2 copies)
London, British Library
(2 copies)
London, Polish Institute Library
London, Royal Astronomical
Society
London, Science Museum
London, University College
Library
London, University of London
Library

London, Victoria and Albert
Museum
London, Dr. William's Library
Manchester, John Rylands
University Library (2 copies)
Oxford, Bodleian Library
Oxford, Christ Church
Oxford, Magdalen College
Oxford, Merton College
Oxford, Wadham College
Oxfordshire, Shirburn Castle
private collection

UNITED KINGDOM (SCOTLAND)
Aberdeen, University Library
(King's College) (2 copies)
Edinburgh, National Library of
Scotland (2 copies)
Edinburgh, Royal Observatory,
Crawford Library
Edinburgh, University Library
Glasgow, University Library
(3 copies)
St. Andrews, University
Library

USA (ARIZONA)
Tucson, University of Arizona

USA (CALIFORNIA)
Beverly Hills, Irwin J. Pincus
private collection
Los Altos, John Warnock
private collection
Los Angeles, Dr. M. N. and
Dr. P. M. Beigelman
private collection

Los Angeles, University of
California, Biomedical
Library
Pasadena, California Institute of
Technology
San Diego, San Diego State
University, Malcolm A.
Love Library
Stanford, Stanford University,
Green Library

USA (CONNECTICUT)
New Haven, Yale Historical
Medical Library
New Haven, Yale University,
Beinecke Library

USA (D.C.)
Washington, Library of
Congress
Washington, Smithsonian
Institution

USA (ILLINOIS)
Chicago, University of Chicago
Library (2 copies)
Urbana-Champaign, University
of Illinois Library
(copy stolen)

USA (INDIANA)
Bloomington, Indiana
University, Lilly Library
Notre Dame, Notre Dame
University Library

USA (KENTUCKY)
Louisville, University of
Louisville Library

USA (MASSACHUSETTS)
Boston, Boston Athenaeum
Boston, Boston Public Library
Cambridge, Harvard University,
Houghton Library (2 copies)
Williamstown, Jay Pasachoff
private collection
Williamstown, Williams
College, Chapin Library

USA (MICHIGAN)
Ann Arbor, University of
Michigan Library

USA (MISSOURI)
Kansas City, Linda Hall Library

USA (NEW JERSEY)
Princeton, Institute for
Advanced Study
Princeton, Princeton University
Library (2 copies)
Princeton, Princeton University,
Scheide Collection

USA (NEW YORK)
Buffalo, Buffalo and Erie
County Public Library
Ithaca, Cornell University
New York, Columbia
University

New York, Morgan Library
New York, New York Public
Library
Schenectady, Dudley
Observatory
West Point, U.S. Military
Academy

USA (OKLAHOMA)
Norman, University of
Oklahoma

USA (PENNSYLVANIA)
Bethlehem, Lehigh University
Haverford, Haverford College
Pittsburgh, Carnegie-Mellon
University, Hunt Library

USA (RHODE ISLAND)
Providence, Brown University,
John Hay Library
Providence, D. G. Siegel
private collection

USA (TEXAS)
Austin, Texas, University of
Texas (2 copies)

VATICAN CITY
Vatican City, Biblioteca
Apostolica Vaticana
(3 copies)

Locations of the Second Edition of *De revolutionibus* (Basel, 1566)

AUSTRALIA
Melbourne, State Library of
Victoria
Sydney, University Library

AUSTRIA
Graz, Universtätsbibliothek
Innsbruck,
Universtätsbibliothek
Melk, Stiftsbibliothek
Vienna, Osterreichische
Nationalbibliothek
Vienna, Universtätssternwarte

BELGIUM
Brussels, Bibliothèque Royale
Albert 1ᵉ
Ghent, Bibliotheek der
Universiteit

CANADA
Montreal, Quebec, McGill
University Library
Ottawa, Ontario, University of
Ottawa, Vanier Library
Toronto, Ontario, University
of Toronto Library

CHINA
Beijing, National Library

CROATIA
Cavtat, Bogišićs Library
Zadar, Naućna Biblioteka

CZECH REPUBLIC
Brno, Augustinian Monastery
Česky Krumlov, Zámecká
Knihovna (Castle Library)
Prague, Národni Knihovna
(Klementinum) (4 copies)
Teplá, Historická Knihovna
Teplá-Kláster

DENMARK
Aarhus, Statsbiblioteket
Copenhagen,
Universitetsbiblioteket

ESTONIA
Tartu, Tartu (Dorpat)
University Library

FRANCE
Ajaccio, Bibliothèque
Municipale
Amiens, Bibliothèque
Municipale
Bordeaux, Bibliothèque
Municipale (2 copies)
Bourges, Bibliothèque
Municipale
Carpentras, Bibliothèque
Municipale Inguimbertine
Clermont-Ferrand,
Bibliothèque Municipale et
Universitaire
Épinal, Bibliothèque
Municipale

Grasse, Bibliothèque
Municipale
Le Mans, Bibliothèque
Municipale
Marseille, Bibliothèque
Municipale
Nantes, Bibliothèque
Municipale
Paris, Bibliothèque de l'Arsenal
Paris, Bibliothèque de
l'Observatoire
Paris, Bibliothèque de la
Sorbonne (2 copies)
Paris, Bibliothèque Mazarine
(3 copies)
Paris, Biblioteka Polska
Paris, Bibliothèque Sainte
Geneviève
Paris, Alain Blanc-Brude
collection
Paris, École Polytechnique
Rouen, Bibliothèque
Municipale
St. Omer, Bibliothèque
Municipale
Toulouse, Bibliothèque
Municipale
Troyes, Bibliothèque
Municipale
Verdun, Bibliothèque
Municipale

GERMANY
Augsburg, Staats- und
Stadtbibliothek

Berlin, Deutsche
Staatsbibliothek
Bonn, Astronomisches Institut
Braunschweig, Technische
Universität
Dillingen a. d. Donau,
Studienbibliothek
Erfurt, Wissenschaftliche
Allgemeinbibliothek
Freiberg, Geschwister Scholl
Gymnasium
Göttingen,
Universitätsbibliothek
Halle, Universitätsbibliothek
Jena, Universitätsbibliothek
(3 copies)
Kiel, Universitätsbibliothek
Mainz, Stadtbibliothek
Minden, Kommunalarchiv
Munich, Bayerische
Staatsbibliothek
Munich, Deutsches Museum
Munich, Universitätsbibliothek
Munich, Universitätssternwarte
Neuburg a.d. Donau, Staatliche
Bibliothek
Nuremberg, Germanisches
Nationalmuseum
Osnabrück, K. Liebmann
private collection
(2 copies)
Stralsund, Stadtarchiv
Trier, Stadtbibliothek
Wittenberg, Evangelisches
Predigerseminar Bibliothek

Wolfenbüttel, Herzog August
Bibliothek
Würzburg,
Universtätsbibliothek
(2 copies)

GREECE
Athens, National Observatory
of Greece

HUNGARY
Budapest, Egyetemi Könyvtár
(University Library)

IRELAND
Dublin, Trinity College
(2 copies)
Maynooth, St. Patrick's College

ITALY
Arezzo, Biblioteca Comunale
Bergamo, Biblioteca Civica
Bologna, Biblioteca Universitaria
Brescia, Universitá Cattolica del
Sacro Cuore
Catania, Biblioteca Regionale
Universitaria
Catania, Biblioteche Riunite
Civica e A. Ursino Recupero
Cremona, Biblioteca Statale
Ferrara, Biblioteca Comunale
Ariostea
Florence, Biblioteca Nazionale
Centrale (2 copies)
Florence, Biblioteca
Riccardiana

Florence, Biblioteca Ximeniana
Florence, Istituto e Museo di
Storia della Scienza
Macerata, Biblioteca Mozzi-
Borgetti
Mantua, Biblioteca Comunale
Milan, Biblioteca Ambrosiana
Milan, Biblioteca Nazionale
Braidense (2 copies)
Milan, Osservatorio
Astronomico di Brera
Naples, Osservatorio
Astronomico di
Capodimonte
Padua, Biblioteca Seminario
Vescovile
Padua, Biblioteca Universitaria
(2 copies)
Padua, Osservatorio
Astronomico
Palermo, Biblioteca Nazionale
Parma, Biblioteca Palatina
Perugia, Biblioteca Comunale
Augusta
Pisa, Biblioteca Universitaria
(2 copies)
Rome, Accademia Nazionale
dei Lincei
Rome, Biblioteca Casanatense
Rome, Biblioteca Nazionale
Centrale (2 copies)
Rome, Biblioteca Universitaria
Alessandrina
Rome, Osservatorio Romano
Siena, Biblioteca Comunale

Treviso, Biblioteca Comunale

Turin, Biblioteca Nazionale
Universitaria

Venice, Biblioteca Nazionale
Marciana

Verona, Biblioteca Capitolare

Vincenza, Giancarlo Beltrame
private collection

JAPAN

Kobe, Sewo Kumoi Collection
for Nofuku-ji Temple
Library

Tokyo (Hino City), Meisei
University

NETHERLANDS

Leiden, Biblioteca Thysiana

Leiden, Universiteitsbibliotheek

POLAND

Cracow, Biblioteka Jagiellońska
(3 copies)

Cracow, Biblioteka Polskiej
Akademii Nauk

Cracow, Muzeum Narodowe
(Czartoryski Library)

Cracow, Muzeum Narodowe
Zbiorny Czapskich

Cracow, University
Astronomical Observatory

Frombork, Muzeum M.
Kopernika

Gdańsk, Biblioteka Gdańska

Jedrzejów, Muzeum im.
Przypkowskich

Kórnik, Biblioteka Kórnicka

Łodż, Biblioteka
Uniwersytecka

Olsztyn, Muzeum Warmii i
Mazur (2 copies)

Poznan, Biblioteka Raczyńskich

Poznan, Biblioteka
Uniwersytecka

Poznan, Pozanańskie
Towarzystwo Przyjaciół
Nauk (2 copies)

Szczecin, Biblioteka Publiczna

Toruń, Biblioteka
Uniwersytecka

Toruń, Książnica Miejska im.
Kopernika

Toruń, Muzeum im. Kopernika

Warsaw, Biblioteka Narodowa
(3 copies)

Warsaw, Biblioteka
Uniwersytecka (6 copies)

Warsaw, Główna Biblioteka
Lekarska (Main Medical
Library)

Wrocław, Biblioteka
Uniwersytecka (2 copies)

Wrocław, Ossolineum
(4 copies)

PORTUGAL

Coimbra, Biblioteca Geral da
Universidade (2 copies)

Lisbon, Biblioteca da Ajuda

Porto, Biblioteca Pública
Municipal

ROMANIA
  Bucharest, Romanian Academy
    of Sciences

RUSSIA
  Moscow, Russian State Library
    (3 copies)
  Moscow, University Library
  St. Petersburg, Academy of
    Sciences Library
  St. Petersburg, National
    Library of Russia
  St. Petersburg, Pulkovo
    Observatory

SLOVAKIA
  Presov, Štátna Vedecká
    Knižnica (State Scientific
    Library)

SPAIN
  El Escorial, Biblioteca de San
    Lorenzo (2 copies)
  Madrid, Biblioteca Nacional
    (2 copies)
  Madrid, Palacio Real
  Salamanca, Biblioteca de
    Universidad (3 copies)
  San Fernando, Observatorio de
    Marina
  Seville, Biblioteca de
    Universidad

SWEDEN
  Djursholm, Institut Mittag-
    Leffler

Stockholm, Kungliga Biblioteket
Stockholm, Kungliga
  Vetenskapsakademiens
  Bibliotek (2 copies)
Uppsala, Astronomiska
  Observatorium

SWITZERLAND
  Basel, Universitätsbibliothek
  Einsiedeln, Stiftsbibliothek
  Geneva, Bibliotheca Bodmeriana
  Geneva, Bibliothèque Publique
    et Universitaire
  Lausanne, Bibliothèque
    Cantonale et Universitaire
  Lucerne, Zentral- und
    Hochschulbibliothek
  Rapperswil, Muzeum Polskie
  Solothurn, Zentralbibliothek
  Zurich, Zentralbibliothek

UKRAINE
  Kharkiv, University Library
  Kiev, National Library of the
    Ukraine
  L'viv, University Library

UNITED KINGDOM (ENGLAND)
  Cambridge, Christ's College
  Cambridge Corpus Christi
    College
  Cambridge, Gonville and Caius
    College
  Cambridge, King's College
    (2 copies)
  Cambridge, Pembroke College

Cambridge, St. John's College

Cambridge, Sidney Sussex College

Cambridge, Trinity College

Cambridge, University Library (2 copies)

Chatsworth, Collection of the Duke of Devonshire

Leeds, Leeds University, Brotherton Library

Lincoln, Lincolnshire Central Reference Library

Liverpool, University of Liverpool

London, British Library

London, J. M. Jedrzejowicz private collection

London, Lincoln's Inn Library

London, London Library

London, Royal Astronomical Society

London, Royal College of Physicians

London, Royal Society

London, Science Museum

London, University of London Library

Manchester, Chetham's Library

Manchester, John Rylands Library

Oxford, Bodleian Library (4 copies)

Oxford, Brasenose College

Oxford, Christ Church

Oxford, Corpus Christi College

Oxford, Hertford College

Oxford, Queens College

Oxford, St. John's College

Oxfordshire, Shirburn Castle private collection

UNITED KINGDOM (SCOTLAND)

Aberdeen, University Library (King's College) (2 copies)

Edinburgh, National Library of Scotland

Edinburgh, Royal Observatory, Crawford Library

Edinburgh, University Library

Glasgow, University Library

St. Andrews, University Library

UNITED KINGDOM (WALES)

Cardiff, National Museum of Wales Library

USA (CALIFORNIA)

Berkeley, University of California, Bancroft Library

Inglewood, Elliott Hinkes private collection

Los Altos, Horace Enea private collection

Los Angeles, Clark Memorial Library

Pasadena, California Institute of Technology, Millikan Memorial Library

Pasadena, Edwin M. Todd
private collection
San Diego, Clay K. Perkins
private collection
San Diego, San Diego State
University, Malcolm A.
Love Library
San Marino, Huntington
Library, Mt. Wilson
Observatory Collection
Stanford, Stanford University,
Green Library

USA (CONNECTICUT)
New Haven, Connecticut, Yale
Historical Medical Library

USA (D.C.)
Washington, Library of
Congress
Washington, Vera Rubin
private collection
Washington, U.S. Naval
Observatory

USA (GEORGIA)
Athens, University of Georgia,
Hargrett Rare Book Library

USA (ILLINOIS)
Chicago, Adler Planetarium
and Astronomy Museum
Chicago, University of Chicago
Library

Urbana-Champaign, University
of Illinois Library

USA (INDIANA)
Bloomington, Indiana
University, Lilly Library

USA (IOWA)
Ames, Iowa State University

USA (KENTUCKY)
Louisville, University of
Louisville Library
Louisville, R. Ted. Steinbock
private collection

USA (MARYLAND)
Baltimore, Johns Hopkins
University Library

USA (MASSACHUSETTS)
Boston, Boston Public Library
Cambridge, Dibner Institute,
Burndy Library (2 copies)
Cambridge, Owen Gingerich
private collection
(2 copies)
Cambridge, Harvard University,
Houghton Library
Williamstown, Williams
College, Chapin Library

USA (MICHIGAN)
Ann Arbor, University of
Michigan Library
Detroit, Public Library

USA (MISSOURI)
Kansas City, Linda Hall Library

USA (NEBRASKA)
Lincoln, M. Eugene Rudd
private collection

USA (NEW JERSEY)
Princeton, Institute for
Advanced Study

USA (NEW YORK)
Ithaca, Cornell University
New York, Brooklyn
Polytechnic
New York, Columbia
University
New York, New York
Academy of Medicine
New York, Frank S. Streeter
private collection
Rochester, University of
Rochester Library
Schenectady, Dudley
Observatory
Syracuse, Syracuse University,
George Arents Research
Library

USA (OKLAHOMA)
Norman, University of
Oklahoma

USA (PENNSYLVANIA)
Bethlehem, Lehigh University
Philadelphia, Library Company
of Philadelphia
Philadelphia, University of
Pennsylvania Library
University Park, Pennsylvania
State University

USA (RHODE ISLAND)
Providence, Brown University,
John Hay Library
Providence, D. G. Siegel
private collection

USA (SOUTH CAROLINA)
Clemson, Clemson University
Library

USA (TEXAS)
Houston, Rice University
Lubbock, Texas Tech
University

USA (VIRGINIA)
Charlottesville, University of
Virginia

VATICAN CITY
Biblioteca Apostolica Vaticana
(3 copies)

Bibliographic Notes

THE TITLE of this adventure is taken ironically from Arthur Koestler's *The Sleepwalkers* (London: Hutchinson, 1959), but actually *The Book Nobody Read* is the story of making my *An Annotated Census of Copernicus' De Revolutionibus (Nuremberg, 1543 and Basel, 1566)* (Leiden: Brill, 2002), a compendium that served as a constant reference in writing the present book. In the process of researching the *Census*, I wrote many essays on various aspects of the history of astronomy, and a number of these essays have been collected in an anthology, *The Eye of Heaven: Ptolemy, Copernicus, Kepler* (New York: American Institute of Physics, 1993). In the notes that follow, I cite some of the essays from the anthology rather than their original publications, which will often be rather more difficult to find.

Four other books were indispensable in writing *The Book Nobody Read*. The first is N. M. Swerdlow and O. Neugebauer's magisterial *Mathematical Astronomy in Copernicus's De Revolutionibus* (New York: Springer, 1984), which includes Swerdlow's concise and authoritative biography of Copernicus. Second, I found myself very frequently checking dates and details in Marian Biskup's comprehensive *Regesta Copernicana (Calendar of Copernicus' Papers)* (Wrocław: Ossolineum, 1973). Edward Rosen's *Three Copernican Treatises* (New York: Octagon, 1971) includes a helpful collection of detailed biographical chapters, and on matters pertaining to Rheticus, Karl Heinz Burmeister's *Georg Joachim Rhetikus, 1514–1574, Eine Bio-Bibliographie* (Wiesbaden: Guido Pressler Verlag, 1967), in three volumes, provided the basic reference source.

For many decades I have written weekly letters to my family; copies of these letters reminded me of many details that would otherwise have faded. In addition, the notes I made of the hundreds of copies of *De revolutionibus* are dated, which proved helpful in solidifying the chronology of these episodes. There are many other places in the text, however, where the curious reader may desire further information or the sources of some of the quotations, and this information is given in the following notes with no attempt to be definitive or comprehensive with respect to the literature.

4: THE LENTEN PRETZEL AND THE EPICYCLES MYTH

Christopher Clavius' remark that Ptolemy's arrangement was not the only way to do it is found in the third edition of his textbook *In Sphaeram Ioannis de Sacro Bosco commentarius* (Rome, 1581), pp. 435–37. Kepler's comment about awakening from sleep is from the beginning of chapter 56 of his *Astronomia nova* (Prague, 1609), and his plaint about making at least seventy tries is in the middle of chapter 16.

My project beloved of the computer magazines was described in *American Scientist*, "The Computer versus Kepler," and is reprinted in *The Eye of Heaven*, pp. 357–66. The follow-up paper seven years later is "The Computer versus Kepler Revisited," in *The Eye of Heaven*, pp. 367–78.

Copernicus' comparison of the Ptolemaic system to a monster is found in the middle of his Preface and Dedication to Pope Paul III. Three English translations of *De revolutionibus* have been made. The first, by Charles Glenn Wallis, appears in *The Great Books of the Western World* (Chicago: Encyclopædia Brittannica, 1952). The second, by Alistair M. Duncan, was published as *Copernicus: On the Revolutions of the Heavenly Spheres* (New York: Barnes and Noble, 1976). The third, an extended project by Edward Rosen, was intended as volume 2 of *Nicholas Copernicus Complete Works* but was published unnumbered in a matching format as *Nicholas Copernicus: On the Revolutions* (Baltimore: Johns Hopkins University Press and London: Macmillan, 1978).

With regard to the epicycles-on-epicycles mythology, my first foray in print on this venerable misconception was "Crisis versus Aesthetic in the Copernican Revolution," reprinted in *The Eye of Heaven*, pp. 193–204. Melvin Tucker and I published "The Astronomical Dating of Skelton's *Garland of Laurel*" in the *Huntington Library Quarterly* 32 (1969): 207–20.

5: "EMBELLISHED BY A DISTINGUISHED MAN"

Astronomical details of the Tower of the Winds are found in Juan Casanovas' "The Vatican Tower of the Winds and the Calendar Reform," in G. V. Coyne, M. A. Hoskin, and O. Pedersen, eds., *Gregorian Reform of the*

Calendar, 1582–1982 (Vatican City: Specula Vaticana, 1983), pp. 189–98 and color pictures of the frescoes are in Fabrizio Mancinelli and Juan Casanovas, *La Torre dei venti in Vaticano* (Vatican City: Liberia Editrice Vaticana, c. 1980).

Tycho's letter to Peucer, 13 September 1588, is found in *Tychonis Brahe Opera omnia*, vol. 7 (Copenhagen, 1924), pp. 127–41. My invited discourse for the International Astronomical Union's Extraordinary General Assembly in Warsaw is titled "The Astronomy and Cosmology of Copernicus" and appears in *The Eye of Heaven*, pp. 162–84.

See Samuel B. Hand and Arthur S. Kunin, "Nicholas Copernicus and the Inception of Bread-Buttering," *Journal of the American Medical Association* 214, no. 13 (1970): 2312–15.

Ernst Zinner's helpful list of seventy locations for the first edition of *De revolutionibus* is found in appendix E to his *Entstehung und Ausbreitung der coppernicanischen Lehre* (Erlangen: Sitzungsberichte der Physikalisch-medizinischen Sozietät zu Erlangen, vol. 74, 1943); Petrus Saxonius' list of his library, which contained many volumes from his teacher, Johannes Praetorius, is reprinted in appendix D.

Robert S. Westman presented two particularly influential papers during the quinquecentennial year, both published two years later: "The Melanchthon Circle, Rheticus, and the Wittenberg Interpretation of the Copernican Theory," *Isis* 66 (1975): 165–93, and "Three Responses to the Copernican Theory: Johannes Praetorius, Tycho Brahe and Michael Maestlin," in Westman, ed., *The Copernican Achievement* (Berkeley and Los Angeles: University of California Press, 1975), pp. 285–345.

6: THE MOMENT OF TRUTH

The results of R. Taton and M. Cazenave's search for Copernican copies in France appear in "Le *De Revolutionibus* en France," *Revue d'Histoire des Sciences* 27 (1974): 318. Coryate's travels and the fanciful picture of the unicorn are found in *Thomas Coriate traveller for the English wits: greeting: from the court of the Grand Mogul, resident at the towne of Asmere, in Easterne India* ([London], 1616).

The closing quotation from Johannes Kepler is paraphrased from the very end of chapter 55 of his *Astronomia nova* (Prague, 1609).

7: THE WITTICH CONNECTION

The remark by Anthony à Wood, published more than a century after his death, is in *Atheniae Oxoniensis* (London, 1815), pp. 491–92. J. L. E. Dreyer's comment on Wittich concludes his paper "On Tycho Brahe's Manual of Trigonometry," *Observatory* 39 (1916): 127–31.

The monograph on Paul Wittich by Owen Gingerich and Robert S. Westman, *The Wittich Connection: Priority and Conflict in Late Sixteenth-Century Cosmology,* is published as *Transactions of the American Philosophical Society* 78, no. 7, (1988). Our "A Reattribution of the Tychonic Annotations in Copies of Copernicus' *De revolutionibus*" appears in *Journal for the History of Astronomy* 12 (1981): 53–54. Owen Gingerich and Miriam Gingerich, "Matriculation Ages in Sixteenth-Century Wittenberg," is in *History of Universities 6* (1987), 135–37.

Edward Rosen's "Render Not unto Tycho That Which Is Not Brahe's" is in *Sky and Telescope 66* (1981): 476–77. His "Was Copernicus' Revolutions Annotated by Tycho Brahe?" is in *Papers of the Bibliographical Society of America 75* (1981): 401–12, and my rejoinder, "Wittich's Annotations of Copernicus," is in *Papers of the Bibliographical Society of America 76* (1982): 473–78.

8: BIGGER BOOKS LINGER LONGER

For additional details on Ursus's attack on Tycho, see Nicholas Jardine, *The Birth of History and Philosophy of Science: Kepler's A Defence of Tycho against Ursus with Essays on Its Provenance and Significance* (Cambridge: Cambridge University Press, 1984).

Information on the editions and surviving copies of Thomas Digges's *A Prognostication Euerlasting* is found under almanacs in A. W. Pollard and G. R. Redgrave, second edition completed by Katharine F. Pantzer, *A Short-Title Catalogue of Books Printed in England, Scotland, and Ireland and of English Books Printed Abroad, 1475–1640* (London: Bibliographical Society, 1986). Information about the piracy of the ship with Pinelli's library on board is found in a note in Agnès Bresson, ed., *Lettres à Claude Saumaise et à son entourage: 1620–1637* (Florence: L. S. Olschki, 1992), pp. 224–26.

For a clear and authoritative account of early printing practices, consult Philip Gaskell, *A New Introduction to Bibliography* (Oxford: Oxford University Press, 1972). Rich details of the Plantin-Moretus Press are found in Leon Voet, *The Golden Compasses: A History and Evaluation of the Printing and Publishing Activities of the Officina Plantiniana at Antwerp* (Amsterdam: Van Gendt, 1969–c. 1972).

A charming, largely anecdotal account of the problems of survival is William Blades, *The Enemies of Books* (London: Trübner, 1879). The quotation from John Bale is in his preface to John Leyland, *The laboryouse journey & serche . . . for Englandes antiquitees . . . by J. Bale* (1549) and is quoted in part in Francis Wormald and G. E. Wright, ed., *The English Library before 1700* (London: University of London, the Athlone Press, 1958), p. 156. The facsimile of the Frankfurt Book Fair catalogs by Georg Willer is published as *Die Messkataloge Georg Willers* (Hildesheim and New York: Olms, 1972–).

9: FORBIDDEN GAMES

The Luther quotation is from his *Tischreden;* the English translation by Theodore G. Tappert, under the general editorship of Helmut T. Lehmann, is *Luther's Works*, vol. 54, *Table Talk* (Philadelphia, Fortress Press, 1967), pp. 358–59. The quotations from Andrew Dickson White, *A History of the Warfare of Science with Theology in Christendom* (New York: Appleton, 1896) are found in vol. 1 on pp. 130 and 127, respectively.

Edward Rosen's detailed detective work on "Calvin's Attitude toward Copernicus" appeared in *Journal of the History of Ideas* 21 (1960): 431–41. A few years later he examined as well "Copernicus on the Phases and the Light of the Planets," in *Organon* 2 (1965): 61–78. Both of these article are reprinted in Rosen's *Copernicus and His Successors* (London and Rio Grande, Ohio: Hambledon Press, 1995), Erna Hilfstein, ed.

The French scholarship cited in the footnote is by Richard Stauffer, "Calvin et Copernic," *Revue de l'Historie des Religions* 179 (1971): 31–40, and a followup discussion by Christopher B. Kaiser is "Calvin, Copernicus, and Castellio," *Calvin Theological Journal* 21 (1986): 5–31. Rheticus' little booklet is transcribed and translated by R. Hooykaas, *G. J. Rheticus' Treatise on Holy Scripture and the Motion of the Earth* (Amsterdam: North Holland, 1984).

For the cosmological impact of the Council of Trent, see Olaf Pedersen, *Galileo and the Council of Trent* (Vatican City: Specola Vaticana, 1991). Details of the Catholic censorship of *De revolutionibus* are found in appendix 3 of my *Census* and also in "The Censorship of Copernicus's *De revolutionibus*," in *The Eye of Heaven*, pp. 269–85.

10: THE HUB OF THE UNIVERSE

My review of the East German facsimile of the princely Peter Apian book appeared as "Apianus' *Astronomicum Caesareum* and Its Leipzig Facsimile," in *Journal for the History of Astronomy* 2 (1971): 168–77.

Concerning celestial circles, see Harold P. Nebelsick, *Circles of God: Theology and Science from the Greeks to Copernicus* (Edinburgh: Scottish Academic Press, 1985).

The Maestlin quotation is from a letter to Kepler, 1 October 1616, in *Johannes Kepler Gesammelte Werke*, vol. 17, no. 744, lines 24–29. The Kepler quotation "Oh ridiculous me!" is from chapter 58 of his *Astronomia nova*; see Kepler's *New Astronomy*, translated by William H. Donahue (Cambridge: Cambridge University Press, 1992). Some details of the effects of Kepler's various improvements of the orbit of Mars are presented in "Giovanni Antonio Magini's 'Keplerian' Tables of 1614 and Their Implications for the Reception of Keplerian Astronomy in the Seventeenth Century" (with James Voelkel), *Journal for the History of Astronomy* 32 (2001): 237–62.

11: THE INVISIBLE COLLEGE

See "The Master of the 1550 Radices: Jofrancus Offusius" (with Jerzy Dobrzycki) in *Journal for the History of Astronomy* 24 (1993): 235–53. For fresh information on Rheticus, I am indebted to the doctoral thesis of Jesse Kraai.

Geographical places of the sixteenth century may be conveniently checked on a CD Rom edition of Gerhardus Mercator's 1595 *Atlas sive cosmographicae meditationes* (Palo Alto: Octavo, 1999).

12: PLANETARY INFLUENCES

For information on Herrera, his library, and the astrology of the Escorial palace, see René Taylor, "Architecture and Magic: Considerations on the

Idea of the Escorial," in Douglas Fraser, Howard Hibbard, and Milton J. Lewine, eds., *Essays in the History of Architecture Presented to Rudolf Wittkower* (London: Phaidon, 1967), pp. 81–109.

The cluster of papers on Galileo's lunar observations and the Cosimo horoscopes include Guglielmo Righini, "New Light on Galileo's Lunar Observations," in Maria Luisa Righini Bonelli and William R. Shea, eds., *Reason, Experiment, and Mysticism in the Scientific Revolution* (New York: Science History Publications, 1975), pp. 59–76, and my response, "Dissertatio cum Professore Righini et Sidereo nuncio," pp. 77–88 in the same volume; also Guglielmo Righini, *Contributo alla interpretazione scientifica dell'opera astronomica di Galileo* (Monografia no. 2, Florence: Instituto e Museo di Storia delle Scienza 1978, no. 2). See also "From *Occhiale* to Printed Page: The Making of Galileo's *Sidereus nuncius*" (with Albert van Helden), *Journal for the History of Astronomy* 34 (2003): 251–67. The quotation from the dedication to Galileo's book is from the translation by Albert Van Helden, *Sidereus Nuncius or The Sidereal Messenger* (Chicago: University of Chicago Press, 1989).

14: THE IRON CURTAIN: BEFORE AND AFTER

Thanks to Robert McCutcheon for the translation of the Deutsch memoir, which appeared in Russian in the *Astronomischeskii Kalendar* for 1953 (Moscow, 1952). F. G. W. Struve's proud description of the library appears in the *Description de l'Observatoire astronomique central de Poulkova* (St. Petersburg, 1845), pp. 237–46. For highlights of the Crawford Library, see *A Heavenly Library: Treasures from the Royal Observatory's Crawford Collection* (Edinburgh: Royal Observatory Edinburgh, and National Museums of Scotland, 1994).

15: PUTTING THE CENSUS TO BED

The splendid Mercator volume is Marcel Watelet, ed., *Gérhard Mercator, Cosmographe le Temps et l'Espace* (Antwerp: Fonds Mercator Paribas, 1994).

Concerning the Putnam thefts in Chicago, consult Jennifer S. Larson, "An Enquiry into the Crerar Library Affair," *AB Bookman's Weekly* 86 (22 January 1990): 280–310.

The other books for which serious attempts have been made to compile a complete census include the Gutenberg Bible, Shakespeare's First Folio, and Audubon's *Birds of America*. See Seymour de Ricci, *Catalogue raisonné des premieres impressions de Mayence (1445-1467)* (Mainz: Gutenberggesellschaft, 1911); Paul Schwenke, *Johannes Gutenbergs zweiundvierezigzeilige Bibel* (Leipzig: Insel-verlag, 1923); Gerhardt Powitz, *Die Frankfurter Gutenberg-Bibel* (Frankfurt: V. Klostermann, c. 1990); Roland Folter, "The Gutenberg Bible in the Antiquarian Book Trade," in Martin Davies, ed., *Incunabula: Studies in Fifteenth-century Books Presented to Lotte Hellinga* (London: British Library, 1999); Sir Sidney Lee, *The First Folio Shakespeare* [New York: *Literary collector*, April 1901]; Anthony James West, *The Shakespeare First Folio: The History of the Book* (Oxford: Oxford University Press, 2001–); Waldemar H. Fries, *The Double Elephant Folio: The Story of Audubon's Birds of America* (Chicago: American Library Association, 1973). Extensive lists of locations of other books have been prepared, generally without attempting to be definitive.

For an engrossing picture of A. S. W. Rosenbach as a wheeler-dealer, see Edwin Wolf 2nd with John F. Fleming, *Rosenbach—A Biography* (Cleveland and New York: World Publishing, 1960).

APPENDIX 1: FROM EQUANT TO EPICYCLET

The relevant papers by Edward S. Kennedy and his students, originally published in *Isis*, are reprinted in *Studies in the Islamic Exact Sciences* (Beirut: American University of Beirut, 1983). A possible avenue for the Muslim devices to have reached Copernicus is discussed in Jerzy Dobrzycki and Richard L. Kremer, "Peurbach and Marāgha Astronomy? The Ephemerides of Johannes Angelus and Their Implications," *Journal for the History of Astronomy* 27 (1996): 187–237.

Index

Note: Page numbers followed by the letter "n" refer to footnotes; page numbers in italics refer to illustrations.